FREEMAN FOX (

The author

Brian Meopham FRICS is General Manager of Contract Advisory Services Ltd and is concerned with the settlement of disputes and claims arising out of construction contracts and with running courses on contract administration for engineers, surveyors and contractors. He has had extensive experience of working in civil and mechanical engineering and construction industries in the UK and overseas.

By the same author
ICE Conditions of Contract—A Commercial Manual
FIDIC Conditions of Contract—A Commercial Manual

Also published by Waterlows
The Layman's Dictionary of English Law *by Gavin McFarlane*
Understanding Business Taxation *by Anthony J. Preston*
Understanding Commercial and Industrial Licensing *by Brendan Fowlston*
Understanding Computer Contracts *by Martin Edwards*
Understanding Dismissal Law *by Martin Edwards*
Understanding Insurance *by Gwilym Williams*
Understanding Recruitment Law *by David Newell*

For further information write to Waterlow Publishers Limited,
Maxwell House, 74 Worship Street, London EC2A 2EN
Tel. 01–377 4600

Extracts from *General Conditions of Government Contracts for Building and Civil Engineering Works* (2nd edition 1977) are Crown copyright and reproduced with the sanction of the Controller, Her Majesty's Stationery Office.

GC/Works/1
Conditions of Contract—
A Commercial Manual

Brian MEOPHAM

WATERLOW PUBLISHERS LIMITED

First edition 1985
© Contract Advisory Services Ltd 1985

Waterlow Publishers Limited
Maxwell House
74 Worship Street
London EC2A 2EN
A member of the British Printing and Communication Corporation PLC

ISBN 0 08 039233 4

British Library Cataloguing in Publication Data

Meopham, Brian
    GC/Works/1 conditions of contract: a commercial manual.—(Waterlow practitioner's library) 1. Building—Contracts and specifications—Great Britain. 2. Public contracts—Great Britain I. Title
    624   TH425

Printed and bound in Great Britain by A. Wheaton & Co Ltd, Exeter

Contents

CHAPTER 1

Some Preliminary Thoughts

This is the second in a series of three books looking at the commercial aspects of conditions of contract used in the building and civil engineering industry. Quite deliberately we are concentrating on other than the JCT forms that have so often been subject to the scrutiny of writers that the shelves of technical libraries must groan under the weight. This book therefore concentrates on the General Conditions of Government Contracts for Building and Civil Engineering Works, more briefly known as Form GC/Works/1.

In the first book of this series, dealing with the ICE Conditions of Contract, we made the point that competent contractual administration is an essential prerequisite to the avoidance of unnecessary claims and loses on construction projects. That message is, if anything, even more relevant in the case of GC/Works/1 Form where administrative matters are left very much more open to the initiative of the user and to 'good practice' than in many other widely used contemporary forms.

Current practice among the drafters of conditions of contract has been to deal with the administrative aspects of running a contract, from the contractual standpoint, very comprehensively in the conditions themselves, leaving little if anything to be covered by separately drafted administrative procedures. Alone among the more modern forms, GC/Works/1 does not do this. As a result it is considerably shorter and on superficial appraisal more easily understood than its contemporaries. This is, however, a trap for the unwary, since so much is left unsaid, presumably on the assumption that the competent administrator would know it anyway, that the administratively inexperienced among contractors and professionals tend to find themselves in problems in abundance.

These conditions endeavour to cater for both building and civil engineering works. In doing so they are perhaps attempting the impossible. Anyone familiar with both the JCT Forms and the ICE Conditions will agree that those respective forms are very different one from the other. The differences are the product of the different problems that are inherent in building and civil engineering work. In seeking to cover both, GC/Works/1 perhaps ends up by covering neither entirely adequately.

1

If it covers the one sector more adequately than the other, it is probably building that comes off best. It is, therefore, perhaps fortunate that much of the government-sponsored civil engineering work undertaken in the UK is actually carried out under the ICE Conditions of Contract and not under GC/Works/1. Certainly one of the largest categories of civil engineering work, the roads programme, is undertaken exclusively under the ICE Conditions. The actual application of the GC/Works/1 conditions to civil engineering work is probably restricted almost entirely to the work of the Ministry of Defence undertaken for the Army, Navy and Air Force.

The next twist in the all-embracing endeavours of the Form is that it tries to cater for all of the various permutations of contract-letting method that might be encountered as well as the varying type of work itself. Thus the conditions endeavour to be drafted to cover lump-sum contracts based on bills of quantities, lump-sums based on specification and drawings, measure-and-value based on schedules of rates, measure-and-value based on bills of approximate or provisional quantities, and any other arrangement that might suit the vagaries of the moment. Perhaps, in trying to be all things for all purposes, the surprise is not that the Form does not always adequately succeed but that it nearly succeeds as often as it does.

Another symptom of the way in which these conditions have not been sufficiently tuned to the needs of the industries that have to work under them is that until recently there was no form of sub-contract published for use in conjunction with the Form as between main contractor and sub-contractor, whether domestic or nominated. Perhaps this is the product of a belief in official circles that the contractor should be responsible for his own destiny and therefore the way in which he lets his sub-contracts, within certain constraints that require the incorporation of some of the clauses of the main contract, is entirely up to him. It is an understandable stance to take, but certainly one that has done nothing to help the smooth administration of government sponsored work under these conditions. That this situation is being remedied is to the benefit of all concerned.

The present form of GC/Works/1 is a development of the earlier Government contract conditions previously known as the CCC/Wks/1 Form. The first edition of these revised conditions was published in November 1973, whilst the second and current edition was published in September 1977, having been further amended to deal more comprehensively with matters of claim, laying down a more adequate basis for their administration. The document in its present form was prepared unilaterally by the Property Services Agency (PSA) without the formal participation of representatives of the construction industry. It is perhaps for that reason that the obligations between the parties are not as evenly balanced as under either the JCT or ICE forms.

However potentially difficult the terms of any contract may appear, even the most impossible can be made to work well when the parties to the bargain are able to act quickly and in the certainty of their authority. Unfortunately such a dynamic approach to the resolution of problems tends to be found in inverse proportion to the size of the organisation dealing with the matter—and therein lies another problem made the more difficult by something known as 'public accountability'. Quick reaction is unlikely to be achieved when people are looking over their shoulder to be sure that they are doing exactly the right thing lest they could be held in any way to blame. It is a sad truism of many large organisations that the less you do, the less likely you are to make a mistake and thereby the better will you prosper. These are not the attitudes, so often the product of an element of risk-taking and dynamic authority exercised with confidence, that foster fast and efficient construction.

Often certain of the professional services on PSA projects are provided by private practices of architects, engineers and quantity surveyors. These organisations are sometimes themselves rather unsure of their authority to act and as a consequence the contractor is liable to find that problem-solving is a two-stage affair with many matters being referred back up the line.

Now compound the problems a little further with 'Departmental policy' in matters of contract interpretation and the fact that there is little available case law to aid this process, and the difficulty of resolving items of dispute quickly can be appreciated.

Matters have therefore in practice to be resolved either at the job level, or if that cannot be achieved, then at quite high levels within the Agency. It is against this background that this book seeks to:

(a) Provide thorough commercial understanding to contractor, engineer, architect and quantity surveyor.
(b) Identify a basis for contract administration that will avoid the main pitfalls of the conditions.
(c) Pin-point procedures that need to be adopted in contract administration.
(d) Ensure that the benefits and protections of the conditions are obtained by both contractor and Agency.

The aim of the book is thus to enhance the profitability of construction companies and assist the smooth running of projects to the benefit of engineer, architect, quantity surveyor, Agency, the Government department client and contractor alike.

CHAPTER 2

Standard letters

In our earlier book on the ICE Conditions of Contract the use of letters related to the actual clauses of the conditions of contract was recommended. The reasoning that lead to that recommendation applies in every way as strongly, if not more so, to the GC/Works/1 conditions of contract.

Most contractors, engineers and architects are familiar with and reasonably well tutored in the obligations and liabilities of the ICE Conditions of Contract and the JCT Forms respectively, dependent on whether their primary area of operation be in civil engineering or building. If one works in civil engineering, contract after contract will be encountered on the ICE Conditions and if in building, on one or other of the JCT Forms. Occasionally those same people and organisations will encounter a Government-sponsored project under the GC/Works/1 Form and their familiarity and expertise in handling projects under those conditions will probably be very much less than in the case of those with which they work day in and day out.

On the other side of the coin, you have staff of the Property Services Agency who may have worked on Government projects for years and to whom the intricacies of GC/Works/1 are second nature. Inevitably this must create an imbalance of contractual expertise to the benefit of the Government and the possible detriment of contractors and practices who only occasionally undertake work for a Government department.

It is for this reason that the standard letter approach is so strongly recommended in the case of GC/Works/1 to avoid the inevitable pitfalls which lack of familiarity must otherwise cause.

Government contracts are no different to any other in that if contractual problems are allowed to accumulate there will inevitably be an adverse effect on profitability, liquidity and even, should matters become too serious, the ability to continue trading. In theory the payment terms of GC/Works/1 are significantly more favourable than those encountered under any of the other standard forms of contract. However, those benefits can all too soon be eroded if backlogs of problems over rating, contentious items of work and claims are allowed to accumulate. In a circumstance in which one side works in a state of comparative unfamiliarity and the other is very concerned with public

4

accountability, the likelihood of such dangerous backlogs developing becomes all too probable.

To the untutored contractor, even the names of the characters in the contract on the Agency's side are unfamiliar. Who or what are the 'Authority' and 'Superintending Officer' with whom the contractor must deal? Even the basic terminology will be unfamiliar to some and is symptomatic of the difficulties that the unwary will find themselves in unless they ensure that their contractual administration is 'right on the ball'.

This chapter therefore sets out standard schedules of administrative correspondence that can be used in conducting the business affairs of the contract, whether as contractor or superintending officer, authority or quantity surveyor, be they employees of the Property Services Agency or staff of professional practices engaged for the particular project.

The letters are listed in two schedules, the first being applicable to the actions of the contractor and the second to those of the authority or superintending officer. Each schedule lists the subject-matter concerned and the clause or clauses in the conditions under which the requirement arises and gives a cross-reference to the chapter in the book where an example of the particular letter may be found (at the end).

In the ICE book we also recommended the use of standard forms by way of various certificates called for under the Conditions. This policy has not been followed in this book, since such stationery either is or may reasonably be expected to be produced by HMSO. If any vacuum exists in this regard, we would not see it as being our role to fill it.

Standard Letter Pro-Formas for use in Conjunction with Actions of the Contractor Under the General Conditions of Government Contracts for Building and Civil Engineering Works Form GC/Works/1 (Edition 2—September 1977)

Reference	*Subject-Matter*	*Contract Clause(s)*	*See Chapter*
GC/1	Request for notification of persons authorised to act for the Authority and their respective powers	1(4)	3
GC/2	Request for sanction to remove items from the site	3(2)	4
GC/3	Notice of additional expense in respect of plant and equipment held on site	3(2), 7(1)(m), 9(2)(a) and (b), 40(5)	4
GC/4	Notification of an error in the bill of quantities	5(2), 9(1)	5
GC/5	Confirmation of possession of the site/order to commence	6	5
GC/6	Non-applicability of bill rates	7(1)(a), 9(1)	6
GC/7	Notice of additional expense arising from a discrepancy	7(1)(b), 9(2)(a) and (b),40(5)	6
GC/8	Notice of additional expense arising from the removal from the site of an item for incorporation	7(1)(c), 9(2)(a) and (b), 40(5)	6
GC/9	Notice of additional expense arising from the removal and/or re-execution of work	7(1)(d), 9(2)(a) and (b), 40(5)	6
GC/10	Notice of additional expense arising from the order of execution of the works or part thereof	7(1)(e), 9(2)(a) and (b), 40(5)	6
GC/11	Notice of additional expense arising from limitations imposed on working hours	7(1)(f), 9(2)(a) and (b), 40(5)	6

Reference	Subject-Matter	Contract Clause(s)	See Chapter
GC/12	Notice of additional expense arising from suspension of the work	7(1)(g), 9(2)(a) and (b), 40(5)	6
GC/13	Notice of additional expense of opening up the work for inspection	7(1)(i), 9(1)(d), 9(2)(a) and (b), 22, 40(5)	6
GC/14	Clarification of the burden of cost for a defect	7(1)(j), 9(1)(d) 9(2)(a) and (b), 32, 40(5)	6
GC/15	Notice of additional expense arising from the execution of emergency work	7(1)(k), 9(2)(a) and (b), 40(5)	6
GC/15	Notice of additional expense arising from the use of materials obtained from excavations	7(1)(1), 9(2)(a) and (b), 20, 40(5)	6
GC/17	Confirmation of an oral instruction	7(2), 7(4), 9(2)(b)(i)	6
GC/18	Application for payment on account of additional expense	40(5)	7
GC/19	Request for the reimbursement of costs in correcting faulty setting out attributable to inaccurate information provided	12, 7(1)(b), 9(2)(a) and (b), 40(5)	8
GC/20	Application for the cost of tests	13(3)	8
GC/21	Request for written confirmation of an instruction of the Resident Engineer/Clerk of Works	16	9
GC/22	Notification of the discovery of antiquities	20(2)	9

Reference	Subject-Matter	Contract Clause(s)	See Chapter
GC/23	Notice of additional expense arising from the discovery of an antiquity	7(1)(m), 20(3), 9(2)(a) and (b), 40(5)	9
GC/24	Notification of the availability of work for examination and/or measurement before covering up	21, 22	10
GC/25	Clarification of the occasions on which the S.O. will require notice prior to covering up work	22	10
GC/26	Notification of additional expense arising from the suspension of work due to inclement weather	9(2)(a) and (b), 23, 40(5)	10
GC/27	Notification of dayworks	24	10
GC/28	Submission of daywork records/and/accounts	24	10
GC/29	Notice of additional expense arising from unforeseen precautions in respect of accepted risks	9(2)(a) and (b), 25(1), 40(5)	10
GC/30	Application for reimbursement of reinstatement costs	26(2)(b)	10
GC/31	Application for extension of time	28(2)	11
GC/32	Request for sanction to sub-let	30(1)	12
GC/33	Objection to the nomination of a Nominated Sub-Contractor/ Nominated Supplier	31(1)	12
GC/34	Clarification of the burden of costs for defects	32	12
GC/35	Request for a 'Schedule of Defects'	32	12
GC/36	Labour return	34	13

Reference	Subject-Matter	Contract Clause(s)	See Chapter
GC/37	Submission of documents in connection with the Final Account	37(2)	13
GC/38	Request for an instruction relating to provisional sum/ provisional quantities	39	14
GC/39	Claim for monthly payment on account	40(3)	15
GC/40	Claim for interim payment between monthly claims	40(3)	15
GC/41	Submission of proof of payment to a nominated sub-contractor/ nominated supplier	40(6)	15
GC/42	Application for payment of the first half of the retention money	41(1)	15
GC/43	Application for further payment after release of the first half of the retention money	41(1) and (4)	15
GC/44	Application for final payment	41(3)	15
GC/45	Request for a certificate of completion	42(1)	15
GC/46	Request for a certificate of satisfactory state	42(1)	15
GC/47	Notification of contributory liability of the Authority in relation to third party liability	47(3)	17
GC/48	Notification in respect of claim for damage to the highway/a bridge	48(4)	17
GC/49	Notice of loss and expense associated with the works of an independent contractor	50, 53	18
GC/50	Notice of prolongation and disruption expenses	53	18

Reference	Subject-Matter	Contract Clause(s)	See Chapter
GC/51	Request for additional information	53(3)(b)	18
GC/52	Request for permission to take progress photographs of the works	58	19
GC/53	Request to concur in the appointment of an arbitrator	61(1)	19

Standard Letter Pro-Formas for use in Conjunction with Actions of the Superintending Officer under the General Conditions of Government Contracts for Building and Civil Engineering Works Form GC/Works/1 Edition 2—September 1977

Reference	Subject-Matter	Contract Clause(s)	See Chapter
GC/SO/1	Delegation of authority-authorised persons	1(4)	3
GC/SO/2	SO approval of goods and materials for the purpose of vesting and becoming the absolute property of the Authority	3(1)	4
GC/SO/3	Consent to the removal of plant/materials from the site	3(2)	4
GC/SO/4	Confirmation of the supply of documents free of charge	4(3)	5
GC/SO/5	Confirmation of the supply of additional copies of documents	4(3)	5
GC/SO/6	Confirmation of the issue of further or modified drawings	4(3)	5
GC/SO/7	Request for the return of all copies of the specification, bills of quantities and drawings	4(4)	5
GC/SO/8	Correction of an error in the bill of quantities	5(2)	5

Reference	Subject-Matter	Contract Clause(s)	See Chapter
GC/SO/9	Notification of the date for commencement of the works	6	5
GC/SO/10	Issue of a superintending officer's instruction in amplification of the contract documents	7(1)	6
GC/SO/11	Notification of rate for varied work	7(1)	6
GC/SO/12	Response to a request for clarification of ambiguity	7(1)(b)	6
GC/SO/13	Instruction to provide sample	7(1)(m)	6
GC/SO/14	Confirmation of oral instruction	7(2)	6
GC/SO/15	Notification to contractor to comply with an instruction of the SO	8	6
GC/SO/16	Request to rectify setting out error	12	8
GC/SO/17	Instruction to carry out test	13(3)	8
GC/SO/18	Order for the removal of materials not in accordance with the contract	7(1)(c), 13(4)	8
GC/SO/19	Order for the re-execution of work not in accordance with the contract	7(1)(d), 13(4)	8
GC/SO/20	Instructions issued to ensure compliance with regulations	14	8
GC/SO/21	Reimbursement of fees paid in respect of royalties and patent rights	15	8
GC/SO/22	Notice of the functions of an assistant of the superintending officer	16	9

Reference	Subject-Matter	Contract Clause(s)	See Chapter
GC/SO/23	Response to a notice of dissatisfaction with an instruction of an assistant of the superintending officer	16	9
GC/SO/24	Notification of safety and/or security requirement	17	9
GC/SO/25	Notice of nuisance/inconvenience to others	18	9
GC/SO/26	Instruction pertaining to discovery of antiquities	20(3)	9
GC/SO/27	Approval/Disapproval of construction excavation	21	10
GC/SO/28	Procedure to be adopted prior to covering up work	22	10
GC/SO/29	Availability for inspection	22	10
GC/SO/30	Instruction to uncover work	22	10
GC/SO/31	Suspension of work due to inclement weather	23	10
GC/SO/32	Lifting of suspension of work	23	10
GC/SO/33	Daily record of resources used on daywork	24	10
GC/SO/34	Daywork sheets	24	10
GC/SO/35	Notification to comply with precautions against loss, damage etc.	25(1)	10
GC/SO/36	Extension of time for the works/ a section of the works during the contract period/at the due date or the extended date for completion	28(2)(i)	11
GC/SO/37	Notification that the contractor is not entitled to [further] extension of time during the contract period/at the due date or extended date of completion	28(2)	11

Reference	Subject-Matter	Contract Clause(s)	See Chapter
GC/SO/38	Notification of intention to deduct liquidated damages	29(1)	11
GC/SO/39	Approval of request for sanction to sub-let	30(1)	12
GC/SO/40	Denial of request for sanction to sub-let	30(1)	12
GC/SO/41	Notification of appointment of nominated sub-contractor/ nominated supplier	31(2)	12
GC/SO/42	Issue of schedule of defects	32	12
GC/SO/43	Request for returns of labour, plant and related matters	34	13
GC/SO/44	Superintending officer's request for replacement of contractor's employee	36(1)	13
GC/SO/45	Authority's request for replacement of contractor's employee	36(2)	13
GC/SO/46	Notification to attend for measurement	37(1)	13
GC/SO/47	Interim/Final payment certificate	40(3)	15
GC/SO/48	Interim payment between monthly claims	40(3)	15
GC/SO/49	Proof of payment to nominated sub-contractor	40(6)(b)	15
GC/SO/50	Final account	41(2)	15
GC/SO/51	Certificate of completion	42(1)	15
GC/SO/52	Certificate of satisfactory state	42(1)	15
GC/SO/53	Special powers of determination	44(2)	16

Reference	*Subject-Matter*	*Contract Clause(s)*	*See Chapter*
GC/SO/54	Determination of contract due to default/bankruptcy of the contractor	45	16
GC/SO/55	Notification of intention to invoke emergency powers	49	17
GC/SO/56	Provision of facilities for other contractors	50	18
GC/SO/57	Acknowledgement/Rebuttal of submission that provision of facilities for other contractors could not reasonably have been foreseen	50	18
GC/SO/58	Refusal of admission to the site	56(1)	19
GC/SO/59	List of contractor's site personnel	56(3)	19
GC/SO/60	Consent to take photographs of the site/works	58	19
GC/SO/61	Notification to concur in the appointment of an arbitrator	61(1)	19

CHAPTER 3

Some Important Definitions

Clause 1(1)—The Contract Documents

'The Contract' means the documents forming the tender and acceptance thereof, together with the documents referred to therein including these Conditions (except as set out in the Abstract of Particulars), the Specification, the Bills of Quantities and the Drawings, and all these documents taken together shall be deemed to form one contract. When there are no Bills of Quantities all references to Bills of Quantities in the Contract shall be treated as cancelled, except that where the context so admits the Schedule of Rates shall be substituted therefor.

Comment

1. It is contemplated that in general there will be at least 7 documents which, taken together, comprise the contract:

(a) The general conditions of contract.

(b) The tender.

(c) The acceptance of the tender.

(d) An abstract of particulars (there is no standard published version of this document).

(e) A specification.

(f) The bills of quantities.

(g) The drawings.

2. It will be noted that both the specification and the bills of quantities are contract documents. Care must be taken to ensure, particularly when the bills of quantities have been prepared in accordance with the Building Method of Measurement, that there is no conflict between the detailed requirements of the bills and the specification. It should be recalled that the Building Method of Measurement does not contemplate a specification being one of the contract documents but that the bill itself should be sufficiently comprehensive to cover all matters of material and workmanship required in the construction of the Works.

In civil engineering works, when bills have been prepared in accordance with the Civil Engineering Standard Method of Measurement, this does not normally present a problem, as the bills merely identify items of work whilst the specification is there to amplify matters. Rarely, however, does the PSA use the Civil Engineering Method of Measurement and often relies on the Building Method of

15

Measurement for civil engineering schemes, for which it is inherently not really suitable—again a potential source of problems and ambiguities.

3. The words 'together with the documents referred to therein' should also be noted. These words refer not only to such items as British Standards, Codes of Practice and the like but also to often voluminous standard documents that are incorporated in the contract, the bulk of which are of little relevance to the project concerned but guaranteed to confuse.

4. The flexible way in which these conditions are employed is apparent in the references made to the fact that bills of quantities may not be employed and the possibility of schedules of rates being a contract document.

Clause 1(2)—Definitions

In the Contract the following expressions shall, unless the context otherwise requires, have the meanings hereby respectively assigned to them.

Comment

1. The *'accepted risks'* are risks for the consequence of which the Authority or Government stays responsible. Some of the items are matters for which, under other forms of contract, the contractor would be responsible and have to carry insurance. In particular he is not responsible for the consequences of fire, explosion, storm, lightning, tempest or flood which are normally the very things most likely to cause extensive damage to the Works.

2. The *'Authority'* means the person so designated in the abstract of particulars. The reference to an abstract of particulars relates to a schedule that sets down matters that are particular to the individual contract and in principle is similar to the Appendix to the Form of Tender of the ICE Conditions. The abstract of particulars is not a standard section bound into the conditions (unlike the situation in almost every other form of contract) and is separately drafted for each contract.

For the purpose of the Contract 'the Authority' is the ultimate power short of arbitration and is rather reminiscent of the engineer under the ICE Conditions. We have in this contract the same two stages of authority that we have been used to in the civil engineering conditions with 'the Authority' being the point of ultimate reference and the superintending officer being the person with whom day-to-day dealings would be conducted. In respect of certain matters 'the Authority' has the ultimate power of decision and these matters cannot be referred to arbitration.

3. '*Bills of Quantities*' includes provisional bills of quantities and bills of approximate quantities. The significance of this definition is that any category of bill would rank as the 'Bills of Quantities' for the purpose of the contract. Lest confusion be caused as to the difference between 'provisional' and 'approximate', there is none, and the contract is merely catering for the loose terminology commonly encountered in documentation.

4. '*The Contract Sum*' is defined in such a way as to cover a lump sum subject to variation or total remeasurement. The conditions are sufficiently flexibly drafted to allow for either arrangment.

5. '*The date for completion*' shall be the date set out in or ascertained in accordance with the abstract of particulars. Once more we see the reliance of these Conditions on the abstract of particulars for putting flesh on the skeletal contractual framework. Confusion is often caused in a situation in which 'the date for completion' is set down as a fixed date without the date for possession of site having been named. Effectively the contract period is therefore of indeterminate length. Ideally the detail in the abstract of particulars should name an elapsed period to be counted from the date for commencement of works.

6. '*The Final Account*' means the document prepared by the quantity surveyor showing the calculation of the final sum. These conditions formally place on the quantity surveyor the obligation to prepare the final account, unlike the situation under the ICE Conditions where the obligation rests on the contractor. This arrangement has disadvantages from the contractor's point of view in that the timely production of the account is largely outside the contractor's hands and his financial destiny is very much controlled by the efficiency of an individual or organisation over whom he has absolutely no control.

7. '*The Schedule of Rates*' means the Authority's schedule of rates in the form stated in the tender or the contractor's schedule of rates. Once more we have an example of the extreme flexibility of these conditions, where the possibility of either the Authority having prepared a schedule of rates or the contractor having prepared it are both envisaged.

8. '*The Site*' means the land or place where work is to be executed under the contract and any adjacent land or place which may be alloted or used for the purpose of carrying out the contract. This definition of the site is most significant as regards Clause 40 and payments on account. The contractor's entitlement to payment relates to work executed on the site and, more importantly, to materials reasonably brought on the site.

There is no provision in these conditions for payment for materials off site.

It should be noted that in the case of certain specialist contractors the site can mean a building or foundation in a sufficiently advanced stage of construction to make the installation of the works of the specialist contractor a feasible proposition.

9. '*The SO*' means the person designated as the superintending officer in the abstract of particulars. The SO in the government form of contract fulfils the same role as the architect under the JCT Form or the engineer's representative under the ICE Conditions.

Clause 1(3)—Permanent and Temporary Works

In these Conditions, references to things for incorporation are references to things for incorporation in the Works and references to things not for incorporation are references to things provided for the execution of the Works but not for incorporation therein.

Comment

1. This clause has relevance in relation to interim payment. Where materials are bought on to the site it is only if the materials are for incorporation in the permanent works that they are to be included in the interim valuation and payment.

2. In the case therefore of materials required for planking and strutting, shoring, formwork and temporary supports, on the basis of the letter of the contract no money should be included in interim certificates for such materials on site.

Clause 1(4)—Delegation of Authority

Any decision to be made by the Authority under the Contract may be made by any person or persons authorised to act for him for that purpose and may be made in such manner and on such evidence or information as he or such person or persons shall think fit.

Comment

1. In order that there may be no doubt as to the authority of the various persons acting on behalf of the Authority under the contract, the contractor should require written notification of the names of the various people concerned and the scope of their authority.

2. It should be noted that under these conditions of contract no 'formal' arrangement exists for this type of notification to be given as of right to the contractor.

3. The fact that the clause states 'any decision to be made by the Authority under the Contract may be made by any person or persons authorised' is rather surprising, since in a number of instances matters are stated to be solely the prerogative of the Authority.

STANDARD LETTERS

Reference: **GC/1**

To: The Authority

Dear Sir

Request for Notification of Persons Authorised to Act for the Authority and their Respective Powers

To assist our contract administration, to the mutual benefit of both the Authority and ourselves, we ask to be informed of the names and designated powers of those persons authorised to act for the Authority under the Contract.

This request is made having regard to Clause 1(4) of the Conditions.

Yours faithfully

Contractor Limited

Reference: **GC/SO/1**

To: The Contractor

Dear Sir

Person/Persons Authorised to Act on Behalf of the Authority

[Further to your request dated .... referenced ....] we/We detail below the names and delegated powers of the person/persons authorised to Act on behalf the Authority under this Contract:

....

....

....

....

This notification is given pursuant to Clause 1(4) of the Conditions.

Yours faithfully

Superintending Officer

CHAPTER 4

Contractor's Enquiries and Vesting in the Authority

Clause 2(1)—Contractor to make Enquiries

The Contractor shall be deemed to have satisfied himself as regards existing roads, railways or other means of communication with and access to the site, the contours thereof, the risk of injury or damage to the property adjacent to the Site or the occupiers of such property, the nature of the materials (whether natural or otherwise) to be excavated, the conditions under which the Works will have to be carried out, the supply of and conditions affecting labour, the facilities for obtaining any things whether or not for incorporation and generally to have obtained his own information on all matters affecting the execution of the Works and the prices tendered therefore except information given or referred to in the Bills of Quantities which is required to be given in accordance with the method of measurement expressed in the Bills of Quantities.

Comment

1. Here we have the equivalent of Clause 11 of the ICE Conditions of Contract, under which the employer will so often make an effort to avoid liability in relation to any inadequacy of information provided to the contractor.

2. The contractor is deemed to have satisfied himself as regards the various physical circumstances affecting the works about which information has not been given in the bills of quantities. The contractor is, however, entitled to take as accurate the information and representations made in the bills of quantities given in accordance with the method of measurement that has governed their preparation.

3. Thus, if in accordance with the method of measurement a particular item had been required to be measured in the bills and had not been so measured, it would not be up to the contractor to include for the matter in his price, even though he might fully have appreciated the need for the item concerned. In this regard reference should also be made to Clause 5.

4. This type of clause does not cover enquiry by the contractor as to whether or not the design of the works and the administrative performance likely to be expected of the engaged consultants is going to be adequate or inadequate and to have priced accordingly. The contractor is entitled to price on the supposition that everyone is going to do what they are supposed to do in accordance with the rules that govern the contract.

5. In view of the very different nature of the Civil Engineering Method of Measurement as against the Building Method (with which these conditions are more usually used), the contractor should be very wary should he find himself undertaking a civil engineering project under these conditions but using the Civil Engineering Method of Measurement. In that case, taking a civil engineering unforeseen ground conditions claim as a case in point, the contractor could find himself in a very exposed position with neither a method of measurement to protect him on the one hand nor contract conditions that adequately cover unforeseen ground conditions on the other.

Clause 2(2)—Disclaimer

No claim by the Contractor for additional payment will be allowed on the ground of any misunderstanding or misinterpretation in respect of any such matter nor shall the Contractor be released from any risks or obligations imposed on or undertaken by him under the Contract on any such ground or on the ground that he did not or could not foresee any matter which might affect or have affected the execution of the Works.

Comment

1. In effect this clause is a very broadly drafted disclaiming clause. As such it falls within the compass of Section 3 of the Misrepresentation Act 1967, under which disclaimers are invalid unless they can be shown to be 'fair and reasonable' in the particular case. The onus of proving that it is 'fair and reasonable' rests on the Authority.

2. Another point to be borne in mind is that if there is an ambiguity or discrepancy in the contract documentation and, for example, the contractor priced on the basis of his cheaper alternative assumption, then having regard to the words used a legal doctrine known as the '*contra proferentem* rule' may come into play, under which the party who has created the ambiguity will have it construed against him.

Clause 3—Vesting of the Works and Materials in the Authority

(1) From the commencement to the completion of the Works, the Works and any things (whether or not for incorporation) brought on the Site in connection with the Contract and which are owned by the Contractor or vest in him under any contract shall become the property of and vest in the Authority subject to his right of rejection of any things for incorporation which are not approved but the Authority shall not, subject to the provisions of Condition 26 and 50, be responsible or chargeable for anything lost, stolen, damaged, destroyed or removed from the Site or that shall fail in any way and the Contractor shall be

VISITOR REGISTRATION

Please complete this card in BLOCK CAPITALS and hand in to reception

Initials and Name _____

Job Title _____

Company _____

Address _____

_____

_____

Nature of Business _____

Photarc Surveys Ltd

has pleasure in inviting

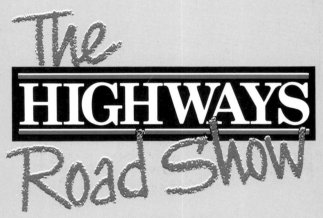

Holiday Inn
Argyle Street, Anderston, Glasgow G3 8RR

GLASGOW 1989

Wednesday, 10th May — 9.30 am to 5.00 pm
Thursday, 11th May — 9.30 am to 4.30 pm

Organised by:
Highways , 111 St James's Road, Croydon,
Surrey CR9 2TH
Tel: 01-684 2660 Tlx: 266332 HET Fax: 01-684 9729

responsible for the protection and preservation of the Works and any things (whether of not for incorporation) brought on the Site until the termination of the Contract.

(2) None of the things referred to in paragraph (1) above which are brought on the Site shall be removed therefrom without the consent in writing of the SO, but the SO may order or permit the Contractor in writing at any time during the progress of the Works to remove from the Site any such things which are unused and thereupon the Contractor shall forthwith remove the same and upon removal the property shall re-vest in the Contractor. The decision of the SO upon any matter arising under this paragraph shall be final and conclusive.

Comment

1. Any material, plant or temporary buildings brought onto the site and owned by or vested in the contractor automatically vest in the ownership of the Authority. The contractor is, however, still liable for any loss which might occur were such items to be stolen, damaged, destroyed or removed from the site.

2. In theory, prior to removal of any plant or equipment or temporary buildings from the site the contractor ought properly to have obtained the SO's sanction in writing to such removal.

In many ways this clause might be considered unreasonable to the contractor, since it delegates to the SO a basic management responsibility which ought properly to rest only in the hands of the contractor.

Who better than the contractor shall say whether or not in the management of a particular contract or his general business that a particular item of plant ought or ought not to be removed? In the event of the SO being unreasonable in this respect, the contractor could find himself with a considerable burden of resource unnecessarily retained on the site due to the SO's failure to grant permission to the contractor for removal of the items concerned.

If such permission is unreasonably withheld it will doubtless be the subject of a contractor's claim. Should the supervising officer issue instructions requiring the unnecessary retention on site of particular items of plant or equipment under the terms of the wide-ranging powers given him under Clause 7(1)(m), the right of the contractor to reimbursement may be found in Clause 9(2)(a).

STANDARD LETTERS

Reference: **GC/2**

To: The SO

Dear Sir

Request for Sanction to Remove Items from the Site

We request written permission to remove the undermentioned items from the Site no longer required for the Works/surplus to requirements/by reason of variation.

(i) ....

(ii) ....

(iii) ....

This request is made having regard to Clause 3(2) of the Conditions.

Yours faithfully

Contractor Limited

Reference: **GC/3**

To: The SO

Dear Sir

Notice of Additional Expense in Respect of Plant and Equipment Held on Site

We refer to our request dated .... asking for your sanction for the removal of ....
from the site. This/these item[s] of equipment/and/plant is/are in our opinion now surplus to construction requirements.

However, noting your refusal to consent to such removal pursuant to your powers of instruction under Clause 7(1)(m), we consider such instruction to be causing us properly and directly to incur expense beyond that provided for in or reasonably contemplated by the Contract and as soon as reasonably practicable we provide notice pursuant to Clause 9(2)(b)(ii) of the Conditions requiring the reimbursement of such additional expense in accordance with Clause 9(2)(a).

Details of the costs incurred are appended/will be forwarded when available in substantiation of our application to be paid such sums by way of advance under Clause 40(5) of the Conditions.

Yours faithfully

Contractor Limited

Reference: **GC/SO/2**

To: The Contractor

Dear Sir

SO Approval of Goods and Materials for the Purpose of Vesting and Becoming the Absolute Property of the Authority

We hereby confirm our approval of the undernoted goods and materials for the purpose of securing payment under Clause 40(2).

Goods and Materials

....

....

....

Such approval is given pursuant to Clause 3(1) of the Conditions and the said goods and materials now vest in and become the absolute property of the Authority.

Yours faithfully

Superintending Officer

Reference: **GC/SO/3**

To: The Contractor

Dear Sir

Consent to the Removal of Plant/Materials from the site

We refer to your request for permission to remove .... from the site and confirm our consent pursuant to Clause 3(2) of the Conditions.

Yours faithfully

Superintending Officer.

CHAPTER 5

Bills of Quantities, Site Possession and Progress

Clause 4(1)—Precedence of Documents

In the case of discrepancy between these Conditions and the Specification and/or the Bills of Quantities and/or the Drawings, the provisions of these Conditions shall prevail.

Comment

1. In construction contracts the precedence of documents in the event of discrepancy will always present problems. The drafter's solution in this case is one more likely to compound than ease the problem, since the standard document, in the form of the conditions of contract, has been given precedence over the very documents that will have been drawn up particular to the individual contract—on the face of it a manifest absurdity and one likely to lead to all manner of unforeseen problems.

2. When it comes to the rest of the contract documents no order of precedence priority is given. The normal common law rules of precedence would therefore apply in which the particular would take precedence over the general, the typed over the printed, and the written over the typed. These commom law rules seek to identify the true intention of the parties and adopt the policy that the document most uniquely drafted to the particular requirements of the individual contract is obviously the one most likely to represent the true intention of the parties.

3. Thereafter if matters still prove irreconcilable the SO is required to issue an instruction under Clause 7(1)(b) which may in its turn provide a basis for recompense to the contractor under Clause 9(2)(a).

Clause 4(3)—Provision of Documents

The SO shall provide free to the Contractor three copies of the Contract Drawings and of the Specification and of the blank Bills of Quantities, and two copies of all further drawings issued during the progress of the Works. The Contractor shall keep one copy of all Drawings and of the Specification on the Site and the SO or his representative shall at all reasonable times have access to them.

Comment

1. Many contractors will find their administrative arrangements call for considerably more than three copies of the original contract drawings and two copies of any further drawings that may be issued.

2. If, as is likely, this is the case, the contractor must make provision in his tender for the additional copies he may require. Equally should the number of drawings issued become out of all proportion to those that could reasonably have been foreseen, the contractor may wish to pursue a minor item of claim in compensation.

Clause 5(1)—Method of Measurement

The Bills of Quantities shall be deemed to have been prepared in accordance with the principles of the Method of Measurement expressed therein, except where otherwise stated.

Comment

1. It will be noted that the clause does not refer specifically to any particular Standard Method of Measurement but merely requires that the one that has been employed shall be stipulated in the bill.

2. In essence these conditions of contract are only really suitable to be employed with the Building Method of Measurement and, as previously noted, their use with the Civil Engineering Method of Measurement is likely to lead to difficulty, since the two documents are not mutually compatible.

Fortunately much Government-sponsored civil engineering work is not actually carried out under these conditions but employs the ICE Conditions of Contract. The most notable example of this is the roads programme.

However, certain work for the nuclear industry and the armed services, of an essentially civil engineering nature, is carried out using these conditions, often creating needless problems, as no mechanism is laid down in these conditions for the resolution of the problems that commonly occur in civil engineering.

3. The concluding words in the condition 'except where otherwise stated' should be noted. Current practice in respect of all current methods of measurement, where divergence occurs between the Standard Method and the method actually employed in carrying out the measurement, requires specific reference to be made to the divergence or it would be deemed to be an error and call into play the requirements of Clause 5(2).

Clause 5(2)—Errors in the Bills of Quantities

Any error in description or in quantity in the Bills of Quantities or any omission therefrom shall not vitiate the Contract nor release the Contractor from his obligations to execute the whole or any part of the Works according to the Drawings and Specification or from any of his obligations or liabilities under the Contract, but where the Contract Sum is based upon the quantities in the Bills of Quantities the error shall be rectified and the rectification dealt with under Condition 9(1) and the value thereof shall be added to or deducted from the Contract Sum as the case may be.

Provided that there shall be no rectification of any errors, omissions or wrong estimates in the prices inserted by the Contractor in the Bills of Quantities or in his computations therein or calculations thereon.

Comment

1. The policy adopted in these conditions as regards errors in the bills of quantities is that now established under all the standard forms of Contract used in the UK. Essentially if the error has been made on the employer's side of the bargain, the matter will be corrected and treated as though it were a variation, whilst if the error has been created by the contractor during the course of pricing, the contractor will have to stand by the consequences of the error.

2. Contractors should note that this right to correction is a mandatory one.

3. The policy of the PSA is to carry out a detailed technical and arithmetical appraisal of the contractor's pricing on tender receipt. In the course of doing this the arithmetical and notable pricing errors made by the contractor in the course of the preparation of his tender are usually identified.

The bill, after discussion with the contractor, is then corrected and the amount of the correction, whether by way of addition or deduction, adjusted as a percentage on the summary to maintain the validity of the original tender sum. This policy is an extremely sound one providing that the PSA does not become overly pedantic in its checking and start arguing about pennies, in which case the whole arrangement can become very time-consuming.

Clause 5(3)—Limitation of Work

The quantities given in Provisional Bills of Quantities or Bills of Approximate Quantities shall not be held to gauge or limit the amount and description of the Works to be executed by the Contractor.

Comment

1. It should be noted that the supposed authority for unlimited variation in the scope of work does not apply to work carried out against a firm bill of quantities. Furthermore, even in the case of provisional and approximate bills of quantities some implied limit of variation must be introduced at some point

2. Conditions of this type must be viewed with caution. The scope of work addition or omission can hardly continue in perpetuity whilst maintaining the validity of the pricing structure adopted by the contractor in the first place. Taken to the extreme, the clause could apparently mean that a contract based on provisional quantities indicating a scope of work of say £10,000 could expand to £10,000,000 without invalidating the contract. Clearly this could not be the intention of any reasonably minded party.

3. This type of clause may well come within the framework of the Misrepresentation Act 1967. Such a disclaiming clause would only be upheld if it were considered to be reasonable by an arbitrator or court. What would be considered reasonable could only be ascertained by reference to the circumstances of the individual contract.

4. This clause is another example of the unfortunate philosophy often found in Government contracts, where the bill of quantities seems to be regarded as no more than a price list of items entitling the Authority to purchase as much or as little of a particular item at the quoted price as may be thought fit.

These conditions, unlike the ICE Conditions, do not recognise that change in quantities as such may serve to invalidate the prices given by the contractor. This is another reason demonstrating the basic unsuitability of these Conditions for use in connection with large-scale civil engineering work. By all means build a catch pit under them but beware the problems of trying to construct a naval dockyard.

Clause 5A—Limitation of Works Against the Schedule of Rates

The description of work given in the Authority's Schedule of Rates shall not define or limit the work to be executed under the Contract.

Comment

1. This clause seeks to deal with comparable circumstances to those noted under Clause 5(3) where the works are being carried out against a schedule of rates. Associated with the schedule of rates there may have been a general description of the scope of work; for instance, on an Area Term Maintenance Contract for general building work on a series of airfields, it might have been stated that the scope of work would be in the order of £500,000 let over a 2-year period.

2. Were that scope of work to prove to be a radical misrepresentation of the work actually to be carried out, remarks previously made against Clause 5(3) would equally apply.

Clause 5B—Lump Sum Prices

If neither Bills of Quantities nor the Authority's Schedule of Rates are provided in respect of the Works or in respect of any work or things to which Condition 38 applies the Contractor shall, if required by the Authority, supply forthwith to the Authority a full and detailed Schedule of Rates which was properly and reasonably used for calculating the Contract Sum or sub-contract sum.

Comment

1. This Clause covers the situation often encountered by a services sub-contractor working as a nominated sub-contractor under the terms of Clause 38 of the Conditions. It will be a situation rarely encountered by the main contractor in his own right.

2. Having given a lump-sum price for an installation, a services sub-contractor will often be asked to prepare a schedule of rates that will be used for the valuation of variation works under his sub-contract.

It is not clear from this clause whether it is the intention that associated with the schedule of rates there should also be a schedule of quantities where quantity and rate taken together substantiate the lump-sum price for the sub-contract works.

3. In view of no specific reference being made to quantities in the clause, it will probably be in a sub-contractor's best interest not to supply such a quantified schedule but only a schedule of rates without quantities attached.

In this way it may be possible to enhance rates that may be used in adjusting variations of addition. Beware, however, this can be a dangerous practice since it may backfire and the rates be used to value omissions, much to the chagrin and detriment of the contractor.

Clause 6—Progress of the Works

Possession of the Site or the order to commence shall be given to the Contractor by notice and the Contractor shall thereupon commence the execution of the Works and shall proceed with diligence and expedition in regular progression or as may be directed by the SO under Condition 7 so that the whole of the Works shall be completed by the date for completion.

Comment

1. It is to be noted that possession of the Site or the order to commence is to be given to the Contractor by notice. Clause 1(6) states 'any notice to be given under the Contract shall be in writing'. The contractor should therefore always have in his possession written notification of the starting date of the contract.

A source of difficulty sometimes encountered under PSA contracts is

that the date for completion is established actually as a 'date' rather than
as an elapsed period of time from the date for commencement. Unlike
both the JCT and ICE forms, the basis of calculating the contract period
is not as well drafted as it might be. Even when clearly drafted this is a
common source of contractual confusion and, as covered here, a virtual
guarantee of it.

2. The clause underlines two points fundamental to the whole contract:
- (a) that the works shall be completed within the period for their
  completion;
- (b) that the works be carried out with diligence and expedition in
  regular progression.

3. It is to be noted that there is no requirement laid down in these
conditions that the date for commencement shall be within a reasonable
time after the date of acceptance of the tender. In the event of this period
being unreasonably protracted, the contractor should endeavour to treat
the notice for commencement as an instruction issued by the SO under
Clause 7(1)(m) and seek compensation under Clause 9(2)(a).

4. The contractor's fundamental obligations outlined in the clause
carry with them a balancing obligation on the part of the Authority that
he shall do everything necessary to enable the obligations placed upon the
contractor to be fulfilled. In practice this means that the site shall be
properly available and information to enable the work to progress in
regular progression shall be available at the proper time.

5. The term 'regular progression' implies the maintenance of a smooth
work flow with a regular throughput of material and no need for the
labour force to fluctuate wildly either by numbers of men working or by
hours per week that the men are employed.

STANDARD LETTERS

Reference: **GC/4**

To: The SO

Dear Sir

Notification of an Error in the Bill of Quantities

We have noted the following [apparent] errors in description/omissions/ points of divergence from the principles of the Standard Method of Measurement where such divergence has not been specifically stated in respect of a specified item or items in the bill of quantities:

(i) ....

(ii) ....

(iii) ....

and request, pursuant to Clause 5(2) of the Conditions, that this/these be corrected and the value thereof ascertained under Clause 9(1).

Yours faithfully

Contractor Limited

Reference: **GC/5**

To: The SO

Dear Sir

Confirmation of Possession of the Site/Order to Commence

Receipt of written notice giving possession of site/the order to commence is confirmed as of .... pursuant to Clause 6 of the Conditions.

Yours faithfully

Contractor Limited

Reference: **GC/SO/4**

To: The Contractor

Dear Sir

Confirmation of the Supply of Documents Free of Charge

We confirm the issue herewith of three copies free of charge of the various documents listed in the appended schedule pursuant to Clause 4(3) of the Conditions.

Should you require further copies of any of the listed documents these may be supplied, and will be invoiced to you at a price to cover printing.

Yours faithfully

Superintending Officer

Reference: **GC/SO/5**

To: The Contractor

Dear Sir

Confirmation of the Supply of Additional Copies of Documents

We note your [verbal] request [made by ..../under

cover of your letter dated .... reference
...................] for additional copies of the documents detailed on the
appended schedule.

We confirm the issue herewith of the items requested and enclose our
invoice for such additional copies in the sum of £.......................
pursuant to Clause 4(3) of the Conditions.

Yours faithfully

Superintending Officer

Reference: **GC/SO/6**

To: The Contractor

Dear Sir

Confirmation of the Issue of Further or Modified Drawings

We confirm the issue herewith of two copies of modified/and/further
drawings pursuant to Clause 4(3) of the Conditions as listed on the
attached 'Drawing Receipt Form'.

You are asked to detach, date, sign and return the detachable section of
the form acknowledging safe receipt by you of the listed drawings.

Yours faithfully

Superintending Officer

Reference: **GC/SO/7**

To: The Contractor

Dear Sir

Request for the Return of all Copies of the Specification, Bills of Quantities and Drawings

You are hereby requested to return all contract documents as listed on the attached schedule and any other documents of a like nature used in the construction of the works.

This notice is given pursuant to Clause 4(4) of the Conditions.

Yours faithfully

Superintending Officer

Reference: **GC/SO/8**

To: The Contractor

Dear Sir

Correction of an Error in the Bill of Quantities

An error/omission has recently come to our attention/been drawn to our attention by you in the Bill of Quantities in that..............................
....

Such error/omission is acknowledged and in respect of ....
on page .... the Bill of Quantities should now read
.... and such error/omission is corrected pursuant to Clause 5(2) of the Conditions.

Any ensuing change in the value of work actually carried out will be ascertained in accordance with Clause 9(1).

Yours faithfully

Superintending Officer

Reference: **GC/SO/9**

To: The Contractor

Dear Sir

Notification of the Date for Commencement of the Works

We hereby give notice that the date for commencement of the works pursuant to Clause 6 of the Conditions shall be.................................
....

Yours faithfully

Superintending Officer

CHAPTER 6

Superintending Officer's Instructions

Clause 7(1)—SO's Instructions

The Contractor shall carry out and complete the execution of the Works to the satisfaction of the SO who may from time to time issue further drawings, details and/or instructions, directions and explanations (all of which are hereafter referred to as 'the SO's instructions') in regard to [items (a) to (m) below.]

Comment

1. The content of Clause 7 relates solely to the powers bestowed on the SO in the issue of instructions. These powers are extremely wide-ranging and far greater than those encountered in most other forms of contract. In the event of the instructions causing the contractor unforeseen expense beyond that provided for in or reasonably to be contemplated by the contract, his remedies will lie under Clauses 9(2) and 53.

2. The matters on which the SO may issue instructions are as follows:

(a) The variation or modification of the design, quality or quantity of the works or the addition or omission or substitution of any work

This authority enables the SO to issue straightforward variation instructions. Where excessive variation works are placed on the contract varying the work in excess of any reasonably implied limit of variation the comments that have previously been made would apply.

(b) Any discrepancy in or between the Specification and/or Bills of Quantities and/or Drawings

Where in order to clarify a discrepancy an instruction is issued which makes it evident that the bills of quantities contain an error either in description or in quantity, then such error would be corrected under the powers of valuation given under Clause 9. The onus would appear to rest on the contractor to highlight any such error and ensure that the appropriate instruction was issued.

Where, in a competitive tender, a contractor may have elected to choose the most economical of the alternatives highlighted by discrepancy and ambiguity, in obtaining a correction he may have to reveal the make-up of his tender and the logic on which his decision was based.

38

(c) The removal from the Site of any things for incorporation which are brought thereon by the Contractor and the substitution therefor of any other such things

In the event of the SO seeking to issue instructions under this section it should be appreciated that the extent of his authority only covers items for incorporation in the Works. Should he endeavour to extend this authority to the Contractor's plant or site accommodation, he would be exceeding his powers, unless it could be shown that he was acting under the terms of section (m) of this clause. It should be remembered that it is basically up to the contractor to decided how his work shall be organised within the constraints that may have been placed upon him by the contract programme included in the tender documents. If the employer seeks to give directions as to the running of the job beyond any such constraint, although the conditions may give him this authority, any demonstrable additional costs should be recoverable by the contractor under Clause 9(2).

(d) The removal and/or re-execution of any work executed by the Contractor

Where the removal and/or execution arises from defective workmanship or the use of defective material on the part of the contractor, the costs arising will of course be to the contractor's account. Where however such removal or re-execution is due to the issue of a late variation, the value of the work should be assessed on a daywork basis in accordance with Clause 9(1)(d).

(e) The order of execution of the Works or any part thereof

We see here another power that enables the SO to impinge on the fundamental right of the contractor to organise his own work. Where instructions are issued to carry out work in an order different to that which could reasonably have been implied from the contract documents or different to that of the contractor's free choice (no comment having been made on the order in which the works were to be executed in the original documents) and such order involves the contractor in additional loss and expense, then such matter would be the subject of a claim under Clause 9(2).

(f) The hours of working and the extent of overtime or nightwork to be adopted

If the instruction issued by the SO is one that changes the conditions and costs of working reasonably envisaged in the original contract tender, then the additional costs should again be recouped by a claim under Clause 9(2).

It will be appreciated that the assumed working hours on which the tender was originally based will have involved a direct correlation with the premium hours that arise in relation to the basic worked hours and a further direct relationship to the on-cost addition for National Insurance and various direct costs associated with the employment of labour.

Obviously any change due to instructions affecting these basic assumptions will alter standard hourly labour costs and should working hours become excessive, radically lower the average output per worked hour.

From the contractor's standpoint there will therefore be two multipliers effectively increasing the cost of labour, one being the higher incidence of premium time and the other being a fall in actual labour productivity. The combination of these two factors when working at the extreme of 7 day working over 12 hour shifts will double the unit labour costs of production. In pursuing a claim for change in working hours, the fall in productivity and its effect on labour costs should also be taken into account.

(g) The suspension of the execution of the Works or any part thereof

This Clause is reminiscent of Clause 40 of the ICE Conditions of Contract but lacks the protections afforded by that clause as to the length of time that suspension can continue. Since such an instruction is likely to cause the contractor prolongation and disruption, his remedy will lie within Clause 53 rather than Clause 9(2).

(h) The replacement of any foreman or person below that grade employed in connection with the Contract

On the wording of the Conditions it would appear that the SO is able to issue such an instruction without detailing his reasons for so doing. If, in the arbitrary implementation of these powers, it were to have a disruptive effect on labour relations, and the contractor believed that the instruction had been unreasonably issued, remedy should be sought under either Clause 9(2) or Clause 53.

(i) The opening up for inspection of any work covered up

Should opening up be ordered in respect on any works which are believed to be defective and having been opened up it is apparent that such works are defective due to some omission on the part of the Contractor, then any instruction issued under this sub-clause would not carry the right to financial reimbursement. However, if no defective works were uncovered, such opening for inspection should be valued under Clause 9(1)(d) and might also give rise to a claim under Clause 9(2) or 53.

(j) The amending and making good of any defects under Condition 32

Care must always be exercised by a contractor receiving an instruction under this sub-clause that he is not in fact receiving a variation order under the guise of making good defects. If the apparent defect has arisen despite the fact that the works are totally in accordance wih the terms of the specification and drawings, the contractor should ask for an order for the removal of the works under sub-clause (d) and for their re-execution under sub-clause (a).

(k) The execution in an emergency of work necessary for security

Additional costs under a heading of this nature would be recovered under Clause 9(1)(d) and might also give grounds for claim under Clause 9(2) or 53. If however such emergency work related directly to the security and stability of the works themselves and the potential state of insecurity was the direct result of some act or omission on the part of the contractor, no charge could of course be made.

(l) The use of materials obtained from excavations on the Site

Should an assumption involving the use of materials arising from the excavation of the Site not have been an integral part of the specification and original design of the Works, it is likely that such an instruction will entail an appreciable degree of re-rating of the contract works. The most likely circumstances will of course relate to filling material, sand and aggregate. In any of these cases the materials concerned are likely to have been a component part in an appreciable number of unit rates which must needs then be re-rated under Clause 9(1)(c).

(m) Any other matter as to which it is necessary or expedient for the SO to issue instructions, directions or explanations.

Although this category of instructional authority appears to give limitless scope to the SO in the nature of the instructions that he issues, such scope will of course be limited by the common law doctrine that the instruction shall relate to the scope of work within the contemplation of the contract.

The SO cannot therefore instruct the contractor to carry out work, unless the contractor should otherwise agree, that is totally outside the scope of reasonable contemplation for the contract concerned. For example, if a contractor had entered into a contract to build a new runway for the Royal Air Force in East Anglia, he could not be required to build another runway in Wales against the authority of an instruction under sub-clause (m).

Clause 7(2)—Confirmation of Oral Instructions

If any of the SO's instructions issued orally have not been confirmed in writing by him such confirmation shall be given upon reasonable request by the Contractor made within 14 days of the issue of such instructons.

Comment

1. These conditions are somewhat ambiguous as to whether or not the contractor would be entitled to reimbursement for variation work against an unconfirmed oral instruction of the SO. It is quite clear, having regard to Clause 9(2)(b)(i), that when it comes to entitlement in regard to matters of claim arising from an instruction, that it must have been given in writing, but on examination of Clause 7(4) it will be seen that, unlike earlier editions of the Contract, reference to written instructions has been dropped in relation to variations as such.

In this confused situation the contractor is most strongly recommended to work on the assumption that he will only be paid for work against written instructions. From the contractor's standpoint the problem can be resolved by refusing to take action on an oral instruction unless confirmed, so requiring the SO to serve a notice under Clause 8 requiring compliance, at which stage in the process of giving the notice, the order will have been given in writing, since a notice can only be actually conveyed to the contractor (having regard to Clause 1(6)) in writing.

2. Contractors must appreciate that the position as regards the confirmation of oral instructions under GC/Works/1 is totally different to the arrangements that are familiar to them under both the JCT and ICE forms. Under these forms, subject only to slightly differing time scales, if a contractor confirms an oral instruction and the confirmation is not dissented from the oral instruction will be deemed to become a written instruction of either the architect or the engineer as appropriate.

Under these Conditions, however, if the SO gives an oral instruction, and it has not been confirmed in writing by the SO himself, the Contractor can request a written confirmation and if within 14 days of the issue of such verbal instruction it has not been so confirmed, technically it would never have been given.

3. If an oral instruction has been received, the procedure to be adopted by the contractor should be as follows:

(a) No immediate works should be carried out against the instruction unless the SO is prepared to give straightaway a written site instruction confirming the order.

(b) If no such immediate confirmation is given and no unsolicited confirmation has been received within (say) one week after the oral instruction has been issued, the contractor should write to the SO requesting formal confirmation of the instruction given.

(c) Noting that if such a request is reasonably made within 14 days there is a mandatory obligation on the SO to confirm the oral instruction, if he does not do so, the oral instruction will be deemed never to have existed, and on no account therefore should the contractor carry out any work of any sort against it.

4. Should a drawing have been issued that contains changes in the Works a confirmatory note of the issue of such a drawing would itself constitute a valid instruction.

Clause 7(3)—Conclusive Nature of Instructions

The decision of the SO that any such instructions are necessary or expedient shall be final and conclusive and the Contractor shall forthwith comply therewith.

Comment

1. The requirement that the Contractor should forthwith comply with an instruction shall only apply where that instruction has been properly issued, i.e. in writing or confirmed in writing.

2. Although again the clause would appear to reinforce the all-pervading nature of the authority of the SO to issue instructions, as we have previously noted under section (m) of Clause 7(1), such instructions must still relate to the work contemplated in the contract concerned.

Clause 7(4)—No Unauthorised Variation to be Made

The Contractor shall not make any alteration in, addition to or omission from the Works described in the Specification and/or Bills of Quantities and/or shown on the Drawings except in pursuance of the SO's instructions issued in accordance with this Condition and such alterations, additions or omissions shall not invalidate the Contract.

Comment

1. This Condition underlines the point that if the Contractor takes action on an instruction that is not issued in accordance with the Conditions he will be in breach of contract, and as such not only liable not to be paid for the work that he may carry out without valid instruction, but possibly also liable for any damage that might stem from such work having been carried out at all.

GCW1-D

2. Attention is again drawn to the possible ambiguity in this condition; compared with earlier editions of the contract, the word 'written' has been deleted in reference to the 'SO's instructions'.

Again, we emphasise the point that no work should on any account be carried out against an unconfirmed oral instruction.

Clause 8—Failure to Comply with Instructions

If the Contractor, after receipt of a notice from the SO requiring compliance with any of the SO's instructions within a period to be specified in the notice by the SO, fails to comply with such instruction, the Authority may, without prejudice to the exercise of his powers under Condition 45, provide labour and/or any things (whether or not for incorporation), or enter into any contract for the execution of any work which may be necessary to give effect thereto and any additional costs and expenses incurred by the Authority in connection therewith over and above those which would have been incurred had the Contractor complied with such instruction shall be recoverable from the Contractor.

Comment

1. This authority to bring on to the site other resources to carry out work which the contractor refuses to undertake is an authority that will be found in all of the standard forms.

2. Clause 45 refers to the right of the Authority to terminate the contractor's contract for breach of contract, failure to comply with instructions being of course an example of such breach. However, it is clearly far more sensible, in the case of a marginal breach, to introduce additional resources from another source and to surcharge the Contractor accordingly.

3. It is again stressed that from the SO's standpoint prior to being able to exercise power under this clause a written notice requiring compliance must have been given. If there has previously been difficulty in getting written confirmation of an instruction from the SO, such confirmation will automatically be provided by the notice requiring compliance with the instruction and the source of the problem will be removed.

STANDARD LETTERS

Reference: **GC/6**

To: The Quantity Surveyor

Dear Sir

Non-Applicability of Bill Rates

We refer to the instruction of the SO dated .... issued pursuant to Clause 7(1)(a) of the Conditions varying the Works by ....

Having regard to the time of issue/and/content of such instruction we consider the work concerned not to be appropriate to valuation under Clause 9(1)(a) or (b) but that its value should be ascertained by measurement and valuation at fair rates and prices.

In this context we note the wording of the concluding paragraph of Clause 9(1) of the Conditions.

Yours faithfully

Contractor Limited

Reference: **GC/7**

To: The SO

Dear Sir

Notice of Additional Expense Arising from a Discrepancy

We refer to your instruction dated .... issued pursuant to Clause 7(1)(b) of the Conditions relating to a discrepancy in/or/between the Specification and/or Bill of Quantities and/or Drawings in that ....

We consider such instruction to be causing us properly and directly to incur expense beyond that provided for in or reasonably contemplated by the Contract and as soon as reasonably practicable we provide notice pursuant to Clause 9(2)(b)(ii) of the Conditions requiring the reimbursement of such additional expense in accordance with Clause 9(2)(a).

Details of the costs incurred are appended/will be forwarded when available in substantiation of our application to be paid such sums by way of advance under Clause 40(5) of the Conditions.

Yours faithfully

Contractor Limited

Reference: **GC/8**

To: The SO

Dear Sir

Notice of Additional Expense Arising from the Removal from the Site of an Item for Incorporation

We refer to your instruction dated .... issued pursuant to Clause 7(1)(c) of the Conditions relating to the removal from the site of .... properly brought onto the site for incorporation in the works and now no longer required by reason of ....

We consider such instruction to be causing us properly and directly to incur expense beyond that provided for in or reasonably contemplated by the Contract and as soon as reasonably practicable provide notice pursuant to Clause 9(2)(b)(ii) of the Conditions requiring the reimbursement of such additional expense in accordance with Clause 9(2)(a).

Details of the costs incurred are appended/will be forwarded when available in substantiation of our application to be paid such sums by way of advance under Clause 40(5) of the Conditions.

Yours faithfully

Contractor Limited

Reference: **GC/9**

To: The SO

Dear Sir

Notice of Additional Expense Arising from the Removal and/or Re-execution of Work

We refer to your instruction dated .... issued pursuant to Clause 7(1)(d) of the Conditions relating to the removal from the site of properly executed work in the form of.............................
....
.... now no longer required by reason of ....
[and its re-execution as detailed by..].

We consider such instruction to be causing us properly and directly to incur expense beyond that provided for in or reasonably contemplated by the Contract and as soon as reasonably practicable provide notice pursuant to Clause 9(2)(b)(ii) of the Conditions requiring the reimbursement of such additional expense in accordance with Clause 9(2)(a).

Details of the costs incurred are appended/will be forwarded when available in substantiation of our application to be paid such sums by way of advance under Clause 40(5) of the Conditions.

Yours faithfully

Contractor Limited

Reference: **GC/10**

To: The SO

Dear Sir

*Notice of Additional Expense Arising from the Order of the Execution of
the Works or Part Thereof*

We refer to your instruction dated .... issued
pursuant to Clause 7(1)(e) of the Conditions relating to the order of
execution of the works/part of the works in that.............................
.... no such requirement having been stipulated in the contract documents/being a
variation from the working order laid down in the contract documents.

Noting the change to be detrimental to the economic performance of our
works as programmed we consider such instruction to be causing us
properly and directly to incur expense beyond that provided for in or
reasonably contemplated by the Contract and as soon as reasonably
practicable provide notice pursuant to Clause 9(2)(b)(ii) of the
Conditions requiring the reimbursement of such additional expense in
accordance with Clause 9(2)(a).

Details of the costs incurred are appended/will be forwarded when
available in substantiation of our application to be paid such sums by
way of advance under Clause 40(5) of the Conditions.

Yours faithfully

Contractor Limited

Reference: **GC/11**

To: The SO

Dear Sir

*Notice of Additional Expense Arising from Limitations Imposed on
Working Hours*

We refer to your instruction dated .... issued pursuant to Clause 7(1)(f) of the Conditions stipulating the hours of working overtime/night work to be adopted of ...., no such requirement having been stipulated in the contract documents/causing a departure from the all-in labour rate build-up used for the tender/and/a fall in labour productivity due to excessive hours.

We consider such instruction to be causing us properly and directly to incur expense beyond that provided for in or reasonably contemplated by the Contract and as soon as reasonably practicable provide notice pursuant to Clause 9(2)(b)(ii) of the Conditions requiring the reimbursement of such additional expense in accordance with Clause 9(2)(a).

Details of the costs incurred are appended/will be forwarded when available in substantiation of our application to be paid such sums by way of advance under Clause 40(5) of the Conditions.

Yours faithfully

Contractor Limited

Reference: **GC/12**

To: The SO

Dear Sir

Notice of Additional Expense Arising from Suspension of the Works

We refer to your instruction dated .... issued pursuant to Clause 7(1)(g) of the Conditions suspending the works/that part of the works being ....

We consider such instruction to be causing us properly and directly to incur expense beyond that provided for in or reasonably contemplated by the Contract and as soon as reasonably practicable provide notice pursuant to Clause 9(2)(b)(ii) of the Conditions requiring the reimbursement of such additional expense in accordance with Clause 9(2)(a).

Details of the costs incurred are appended/will be forwarded when available in substantiation of our application to be paid such sums by way of advance under Clause 40(5) of the Conditions.

Yours faithfully

Contractor Limited

Reference: **GC/13**

To: The SO

Dear Sir

Notice of Additional Expense Arising from Opening Up the Work for Inspection

We refer to your instruction dated .... issued pursuant to Clause 7(1)(i) requiring the opening up for inspection of ....

[This work was the subject of a notification of availability for inspection dated .... issued in accordance with the requirements of Clause 22 but the SO elected not to inspect.]

As the work has been opened up, inspected and found to be totally in accordance with the requirements of the Contract, reimbursement is sought valued in accordance with Clause 9(1)(d).

We further consider such instruction to be causing us properly and directly to incur residual expense beyond that provided for in or reasonably contemplated by the Contract and as soon as reasonably practicable provide notice pursuant to Clause 9(2)(b)(ii) of the Conditions requiring the reimbursement of such additional expense in accordance with Clause 9(2)(a).

Details of the costs incurred are appended/will be forwarded when available in substantiation of our application to be paid such sums by way of advance under Clause 40(5) of the Conditions.

Yours faithfully

Contractor Limited

Reference: **GC/14**

To: The SO

Dear Sir

Clarification of the Burden of Cost for a Defect

We refer to your instruction dated .... issued pursuant to Clause 7(1)(j) of the Conditions requiring the making good of ...., a defect [and enclosing a schedule of defects] purportedly notified under Clause 32 and question whether such item/the following items:

(i) ....

(ii) ....

(iii) ....

is/are attributable to or arise[s] from any failure or neglect on the part of the contractor or any sub-contractor or supplier in the proper performance of the Contract or from frost occurring before completion of the works.

Believing this item/these items of purported defect to have arisen from some other cause, we request their value be ascertained and paid for as if it were additional work under the terms of Clause 9(1)(d) of the Conditions.

We further consider such instruction to be causing us properly and directly to incur residual expense beyond that provided for in or reasonably contemplated by the Contract and as soon as reasonably practicable provide notice pursuant to Clause 9(2)(b)(ii) of the Conditions requiring the reimbursement of such additional expense in accordance with Clause 9(2)(a).

Details of the costs incurred are appended/will be forwarded when available in substantiation of our application to be paid such sums by way of advance under Clause 40(5) of the Conditions.

Yours faithfully

Contractor Limited

Reference: **GC/15**

To: The SO

Dear Sir

Notice of Additional Expense Arising from the Execution of Emergency Work

We refer to your instruction dated .... issued pursuant to Clause 7(1)(k) of the Conditions concerning the execution of .... being emergency work necessary for security. Believing the need for this emergency work to have arisen through no failure or neglect on the part of the Contractor in the proper performance of the Contract, we now consider such instructions to be causing us properly and directly to incur expense beyond that provided for in or reasonably contemplated by the Contract and as soon as reasonably practicable provide notice pursuant to Clause 9(2)(b)(ii) of the Conditions requiring the reimbursement of such additional expense in accordance with Clause 9(2)(a).

Details of the costs incurred are appended/will be forwarded when available in substantiation of our application to be paid such sums by way of advance under Clause 40(5) of the Conditions.

Yours faithfully

Contractor Limited

Reference: **GC/16**

To: The SO

Dear Sir

Notice of Additional Expense Arising from the Use of Materials Obtained from Excavation

We refer to your instruction dated .... issued pursuant to Clause 7(1)(l) of the Conditions requiring the use of sand/aggregates/filling material/hardcore [in the form of ....
..................] obtained from excavations on the Site.

As such excavated materials had been described in the contract bills, so far as disposal was concerned, as 'cart to tip provided by the Contractor', account had been taken of their value and allowed for in the contract price, no stipulation having been made in the tender documents to the contrary.

We therefore consider the saving being ascertained by the quantity surveyor to the Authority's benefit under Clause 20 to be causing us properly and directly to incur expense beyond that provided for in or reasonably contemplated by the contract and as soon as reasonably practicable we provide notice pursuant to Clause 9(2)(b)(ii) of the conditions requiring the reimbursement of such additional expense in accordance with Clause 9(2)(a).

Details of the costs incurred are appended/will be forwarded when available in substantiation of our application to be paid such sums by way of advance under Clause 40(5) of the Conditions.

Yours faithfully

Contractor Limited

Reference: **GC/17**

To: The SO

Dear Sir

Confirmation of an Oral Instruction

In accordance with the requirements of Clause 7(2) of the conditions your written confirmation of the oral instruction issued by .... in respect of .... is requested. This request is made within fourteen days of the issue of such oral instruction.

Until written confirmation is received no action can be taken on the order having regard to the terms of Clauses 7(4) and 9(2)(b)(i) of the conditions.

Yours faithfully

Contractor Limited

Reference: **GC/SO/10**

To: The Contractor

Dear Sir

Issue of a Superintending Officer's Instruction in Amplification of the Contract Documents

We confirm the issue of the attached 'SO's Instruction Form ....'. This instruction is issued pursuant to Clause 7(1) .... of the conditions.

Yours faithfully

Superintending Officer

Reference: **GC/SO/11**

To: The Contractor

Dear Sir

Notification of Rate for Varied Work

We refer to our recent consultations on the subject of a rate/rates for
.... arising as a result of variations
pursuant to Clause 7(1).

[We are pleased to confirm our agreement to the undernoted rate/rates.]

[We regret that we are unable to agree your rate/proposal/proposals and
now determine the rate/rates as noted below.]

Rate	*Item*
....	....
....	....
....	....

In agreeing/determining the rate/rates we have followed the determining
principles of the Contract and now accordingly so notify you pursuant to
Clause 9(1) of the Conditions.

Yours faithfully

Superintending Officer

Reference: **GC/SO/12**

To: The Contractor

Dear Sir

Response to a Request for Clarification of Ambiguity

We refer to your letter dated .... reference .... detailing an apparent discrepancy/ambiguity in the documents forming the Contract.

[We do not accept that there is any discrepancy/ambiguity, since if you examine .... you will observe that .... and you should therefore proceed to construct the works accordingly.]

[We thank you for drawing our attention to the matter [in such a timely way] and agree with you that there does appear to be a(n) ambiguity/discrepancy in the documents.]

[Pursuant to Clause 7(1)(b) of the Conditions we therefore instruct you to ....]

Yours faithfully

Superintending Officer

Reference: **GC/SO/13**

To: The Contractor

Dear Sir

Instruction to Provide Sample

You are directed to provide a sample .... consisting of .... pursuant to Clause 7(1)(m) of the Conditions.

Since such requirement was [not] clearly intended by or provided for in the contract, the cost will be borne by the Authority/yourself.

Yours faithfully

Superintending Officer

Reference: **GC/SO/14**

To: The Contractor

Dear Sir

Confirmation of Oral Instruction

We enclose herewith our 'Confirmation of Oral Instruction', reference no. ....

This confirms the instruction orally given by...................................
.... /pertaining to...
/pursuant to Clause 7(2) of the Conditions.

Yours faithfully

Superintending Officer

Reference: **GC/SO/15**

To: The Contractor

Dear Sir

Notification to Contractor to Comply with an Instruction of the SO

We refer to SO Instruction No. .... dated
.... for
....
.... which requires you to carry out work
pursuant to Clause 7(1) and note that to date no action has been taken by
you.

We hereby give notice as required by Clause 8 of the Conditions that the
Authority will employ and pay other persons/enter into a contract to
execute the work necessary to comply with the above instruction and that
all costs incurred will be recoverable by the Authority as a debt and

deducted from any advances on account that are due or may become due pursuant to Clause 40(1) of the Conditions.

Yours faithfully

Superintending Officer

CHAPTER 7

Valuation of the Superintending Officer's Instructions

Clause 9(1)—Valuation of the SO's Instructions

The value of alterations in, additions to and omissions from the Works made in compliance with the SO's instructions shall be added to or deducted from the Contract Sum, as the case may be, and shall be ascertained by the Quantity Surveyor as follows:

(a) by measurement and valuation at the rates and prices for similar work in the Bills of Quantities or Schedule of Rates in so far as such rates and prices apply;

(b) if such rates and prices do not apply, by measurement and valuation at rates and prices deduced therefrom in so far as it is practicable to do so;

(c) if such rates and prices do not apply and it is not practicable to deduce rates and prices therefrom, by measurement and valuation at fair rates and prices; or

(d) if the value of alterations or additions cannot properly be ascertained by measurement and valuation, by the value of materials used and plant and labour employed thereon in accordance with the basis of charge for daywork described in the Contract.

Provided that where an alteration in or addition to the Works would otherwise fall to be valued under sub-paragraph (a) or (b) above but the Quantity Surveyor is of the opinion that the instruction therefor was issued at such a time or was of such content as to make it unreasonable for the alteration or addition to be so valued, he shall ascertain the value by measurement and valuation at fair rates and prices.

Comment

1. It should be appreciated by those more used to administrative arrangements involving total remeasurement of work under the ICE Conditions of Contract that we are here dealing with a clause drafted against the background of a lump-sum contract subject to variation. If, even under these conditions, the contract had been intended to be subject to total remeasurement, it would have been based on provisional bills of quantities, bills of approximate quantities or a schedule of rates, in which case the appropriate method of valuation would be that laid down in Clause 10.

2. The clause appears to give the quantity surveyor named in the abstract of particulars full powers under the contract to carry out the function of the measurement and valuation of the works, and as we will see later, any expense ascertainment required in connection with claims.

In practice this is a power often curtailed under the terms of the letter of appointment from the Authority to a firm of quantity surveyors. Their appointment will often have been circumscribed to an authority to carry out only the most routine measurement and valuation at Bill rates with all powers calling for the exercise of judicial judgement being retained within the Authority.

Only rarely, if ever, is the contractor actually informed of this and in practice it only becomes apparent as the contractor becomes increasingly frustrated when he finds that it is totally impossible to get anything, other than matters of a most routine nature, resolved with the quantity surveyor, whom he has been led to believe has authority to act under the Conditions. So blatant has this become that adequate recovery is not being made by contractors whilst their right to claim in such circumstances is debarred, as Clause 9(2) does not cover this situation.

The PSA, contrary to their own published conditions, would appear to wish to sweep as many of these contentious items into the area of extra-contractual claims as possible, thereby presumably delaying the final financial settlement as long as possible, and enhancing the liquidity position of the Department.

An awareness by the contractor's representative of this situation may be of value in reaching rational settlement with the appointed quantity surveyors.

3. It is against that background that the mode of ascertainment of the value of variations should be considered.

Although these conditions appear to have carefully avoided using the words 'principles' or 'rules' to describe the methods of valuation, as is the case in other standard forms, the status of the methods of valuation under these conditions would still appear to be that of a rule, having regard to the words 'shall be ascertained as follows'.

The methods are:

(a) 'By measurement and valuation at the rates and prices for similar work in the Bills of Quantities or Schedules of Rates, in so far as such rates and prices apply.'

For most work this will represent a perfectly fair basis of valuation. Where, however, such works are not of a similar character executed under similar conditions to those envisaged in the original contract, or are such as will involve the contractor in loss and expense beyond that contemplated in the original contract, care must be taken to ensure that the works are then valued under either sub-clause (b), (c) or (d) of the Clause.

If this cannot be achieved, the contractor is precluded from obtaining a remedy by making a claim under Clause 9(2)(a), having regard to the wording of Clause 9(1), but he still has a claim remedy under Clause 53.

However, bearing in mind the time normally taken to clear claims with the Authority, it is infinitely to be preferred that matters be resolved through the normal measurement and rating of the works.

(b) 'If such rates and prices do not apply, by measurement and valuation at rates and prices deduced therefrom in so far as it is practicable to do so.'

This sub-clause generally covers the situation in which the agreement of pro-rata rates would be required. The more it is possible for the contractor to build up satisfactory rates offering a reasonable return around some form of pro-rata synthesis against the existing rates in the contract, the less contentious is agreement likely to be and the quicker his recovery.

In the course of measurement the surveyor should pay attention to the terms of the particular method of measurement stipulated in the contract and ensure as far as possible that the requirements of that stipulated method are scrupulously followed.

In this way it may be possible to have work of dissimilar characteristic to that contained in the original contract appropriately measured in accordance with acknowledged rules that reflect this dissimilarity and require a separately identified item to be measured.

Since it has been seen that under Clause 5(1) the bills of quantities are deemed to have been prepared in accordance with a stipulated method of measurement, this approach would appear to be an irrefutable basis on which some reasonable settlement of some contentious items could be obtained.

(c) 'If such rates and prices do not apply and it is not practicable to deduce rates and prices therefrom, by measurement and valuation at fair rates and prices.'

This sub-clause covers the agreement of star rates. The basis for the agreement of star rates should always be against actual costs incurred with the addition of a fair mark-up related generally to the levels established in the original tender.

The more the measurement of variations is related to star rates, the more nearly will the contract be valued on a negotiated basis. The skill applied to proper item differentiation in measurement will lay a firm foundation for the agreement of reasonable rate levels for the work concerned.

(d) 'If the value of alterations or additions cannot be properly ascertained by measurement and valuation, by the value of materials used and plant and labour employed thereon in daywork in accordance with the basis of charge for daywork described in the Contract.'

Where it is apparent that the contractor will wish to have work valued on a star rate basis, it will be a sensible precaution to maintain contemporary daywork records and to argue, if the star rate levels of pricing prove inadequate, that the rates cannot be properly ascertained and that therefore the daywork basis for reimbursement should be used.

It should be appreciated that as a fundamental common law doctrine the contractor should not be worse off by the execution of a variation than he would have been had the variation not been issued. The proviso to that statement is however that should the contractor have a bad rate in his contract the extension of the scope of that work against a bad rate will still have to be borne by the contractor, providing that it can be shown that the additional work is still of a similar character and executed under similar conditions to that to which the bad rate originally applied.

4. The concluding paragraph of Clause 9(1) makes it clear that the quantity surveyor is entitled to take into account the time at which a variation order was issued when deciding whether or not the application of rates and prices in accordance with sub-clauses (a) or (b) would be reasonable. Should he conclude that it would not be reasonable, he shall then ascertain the value by measurement and valuation at fair rates and prices—i.e. he can resort to the mode of rating established by sub-clause (c).

Clause 9(2)—Claims Other Than for Prolongation and Disruption Expenses

(a) If the Contractor:

    (i)   properly and directly incurs any expense beyond that otherwise provided for in or reasonably contemplated by the Contract in complying with any of the SO's instructions (other than instructions for alterations in, additions to or omission from the Works), or

    (ii)   can reasonably effect any saving in the cost of the execution of the Works in or as a result of complying with any of the SO's instructions,

the Contract Sum shall, subject to sub-paragraph (b) of this paragraph and to Condition 23, be increased by the amount of that expense, or shall be decreased by the amount of that saving, as ascertained by the Quantity Surveyor.

(b) It shall be a condition precedent to the Contract Sum being increased under sub-paragraph (a) of this paragraph that:

    (i)   the SO's instructions shall have been given or confirmed in writing;

(ii)   as soon as reasonably practicable after incurring the expense the Contractor shall have provided such documents and information in respect of the expense as he is required to provide under Condition 37(2); and

(iii)   the instructions shall not have been rendered necessary as a result of any default on the part of the Contractor.

Comment

1. This clause deals with the recovery of additional expense incurred by the contractor otherwise than resulting from a variation order (i.e. an SO's instruction issued under Clause 7(1)(a)), or from a matter involving prolongation and disruption, in which case the claim would be lodged under Clause 53.

2. As an example of the type of circumstance falling for recovery of additional expense under this clause, an instruction issued by the SO under Clause 7(1)(f) altering the working hours would be a typical case in point. It might very well not cause the contractor disruption and might indeed even result in a shortening of the contract period; however due to the combination of an unforeseen level of premium hours and a possible fall in productivity due to extended hours, the contractor would clearly incur additional expense beyond that otherwise provided for in, or reasonably contemplated by, the contract.

Turn then to Clause 9(2)(a).

3. The clause also contemplates the possibility of an instruction of the SO causing a saving in cost to the contractor, in which case the Authority would look to a saving.

4. If the contractor is to seek any recovery under this clause, a written instruction from the SO must exist. The contractor must provide the quantity surveyor with all the documents and information necessary to make a calculation of the additional expense, whilst the need for the instruction must not have been the product of some default on the part of the contractor.

As an example, if the contractor had delayed the works through his own default and timely completion was a matter of extreme emergency, the SO would be entitled to issue an instruction to adopt additional working hours and the contractor would not then be entitled to compensation for the additional costs so caused him.

5. In the November 1973 first edition of the conditions at this point, in Clause 9(4), provision was made for interim payment on account. Although the right has been deleted here, it has in fact been incorporated under the advances on account in Clause 40(5), under which if the SO is of the opinion that there is an increase or decrease, he can decide an amount to be added or deducted on account of that increase or decrease.

6. The serving of a notice by the contractor under this clause is not a condition precedent to his right to recovery, although should he need to seek recovery under Clause 53, the service of a notice in relation to that Clause *is* a condition precedent.

It is, however, recommended that the contractor serve a notice in relation to this clause even though it is not specifically called for.

Clause 9(3)—Unilateral Ascertainment of Value by the Quantity Surveyor

If any alterations or additions (other than those authorised to be executed by daywork) have been covered up by the Contractor without his having given notice in pursuance of the provisions of Condition 22 of his intention to do so the Quantity Surveyor shall be entitled to appraise the value thereof and his decision shall be final and conclusive.

Comment

1. This clause will particularly relate to underground works where the contractor has covered up the work without giving the SO an opportunity to inspect it and presumably to require the quantity surveyor to ascertain the value, or be a party to the ascertainment of its value by measurement.

2. It still seems reasonable to argue that there must be an implied condition that the valuation placed upon the matter by the quantity surveyor must be fair and reasonable, even though it is stated that his decision thereon shall be final and conclusive.

Such decision cannot be taken arbitrarily and unprofessionally in an unreasonable way at a value such that a reasonable and experienced contractor or surveyor would never have concluded to be the likely and true measurement and valuation of the work concerned.

Clause 10—Remeasurement Contracts

When the Contract is based upon Provisional Bills of Quantities, Bills of Approximate Quantities or the Schedule of Rates the value of the whole of the work executed to the satisfaction of the SO shall be ascertained by measurement and valuation in accordance with Condition 9(1) (except as may otherwise be provided in respect of any item to which Condition 38 applies).

Comment

1. This clause applies to those contracts which are to be the subject of total remeasurement. Variations will not be identified as such but will be taken into the bulk measurement of the whole of the works in much the same way as is the normal procedure under a civil engineering Contract.

2. Within such a contract there might still be nominated sub-contractors under the provision of Condition 38 working on a lump sum or some other basis.

Clause 11G—Variation of Price for Labour-Tax Matters

In view of the changes that have taken place to a formula-adjustment basis in regard to fluctuation, the whole clause has not been set out, although the commentary will give adequate information as to the nature of the clause.

Comment

1. In general, this clause enables the contract value to be adjusted at the net rate of any change in a labour tax matter excluding only value added tax, income tax and particularly any levy payable by the employer under the Industrial Training Act 1964.

2. Similar provisions are to be applied to any nominated sub-contract which is entered into by the main contractor.

3. The contractor will be required to maintain adequate records to demonstrate the increase or decrease being experienced. The increase or decrease so ascertained shall be added to or deducted from the contract sum on completion of the contract unless the parties otherwise agree.

In the rare circumstances where the labour tax matter was leading to a saving, it would doubtless be to the contractor's advantage from a cash flow standpoint to request that the adjustment be made on completion. In the more normal circumstances of an increase it will obviously pay him to press the Authority to agree the tax increases in a progressive way as the job proceeds and for the sum to be included in interim certificates.

4. The Clause envisages that the basis of adjustment will be against the man days actually worked on the site, one man day being counted as one fifth of a weekly contribution.

Clearly no provision is made that would allow the Authority to base any adjustment on the number of man days built into the original tender.

STANDARD LETTER

Reference: **GC/18**

To: The Authority

Dear Sir

Application for payment on Account of Additional Expense

We refer to our notice dated .... and reiterate our request contained therein for payment, by way of advance under Clause 40(5) of the Conditions, of our properly and directly incurred additional expense.

Yours faithfully

Contractor Limited

CHAPTER 8

Setting Out, Compliance with Specification, Tests and Fees

Clause 12—Setting out the Works

The SO shall supply dimensioned drawings, levels and other information necessary to enable the Contractor to set out the Works. The Contractor shall set out the Works in accordance therewith and shall provide all necessary instruments, profiles, templates and rods and be solely responsible for the correctness of such setting out. The Contractor shall provide, fix and be responsible for the maintenance of all stakes, templates, profiles, level marks, points and such other setting out apparatus as may be required, and shall take all necessary precautions to prevent their removal, alteration or disturbance and shall be responsible for the consequences of such removal, alteration or disturbance and for their efficient reinstatement.

Comment

1. In practice, if the nature of the works is fairly simple, no problem is likely to be encountered in setting out. The very simplicity will ensure that the information provided is probably accurate and that the problems associated with physically marking out the works on the ground will be of such minor nature as to be unlikely to lead to mistake. All is not, however, so simple in the case of complex and extensive works where mistakes made in the setting-out information provided to the contractor, particularly where it may rely on control points attached to a detailed land survey, may have serious effects on cost and delay to the contract.

2. There is no onus placed on the contractor to check the information provided for setting-out purposes, since as will be seen from the clause, 'the Contractor shall set out the works in accordance therewith'. Where a discrepancy subsequently comes to light, it should be corrected by way of an instruction issued under Clause 7(1)(b) and the contractor accordingly recompensed for any additional work undertaken under Clause 9(1) and any consequential unrecovered costs under Clauses 9(2) or even 53, should it cause prolongation and disruption.

Clause 13(1)—Materials to Comply with Specification

All things for incorporation shall be of the respective kinds described in the Specification and/or Bills of Quantities and/or Drawings and the Contractor shall upon request of the SO prove to the SO's satisfaction that such things do so conform.

Comment

The words 'all things for incorporation' are only an obstruse way of saying 'materials'. The contractor's difficulty in ensuring that his price initially covers the specified materials, and subsequently the execution of the Works, and that he obtains the appropriate materials, stems from the tendency of PSA Government contracts to contain a startling plethora of standard specifications, of which about 90% have no application to the particular works forming the contract and where often the significant 10% become obscured and unappreciated in a welter of irrelevant paper.

Clause 13(2)—Right of Inspection

The SO and his representative shall have power at any time to inspect and examine any part of the Works or anything for incorporation either on the Site or at any factory or workshop or other place where any such part or thing is being constructed or manufactured or at any place where it is lying or from which it is being obtained, and the Contractor shall give all such facilities as the SO or his representative may require to be given for such inspection and examination.

Comment

This requirement giving free access for the inspection of the works together with an access arrangement to manufacturing premises is a standard requirement in all construction contracts. Since most construction materials are purchased by a contractor from organisations outside his own group, it is essential that he incorporates such access requirements in any purchase order.

Clause 13(3)—Tests

The SO shall be entitled to have tests made on any things for incorporation supplied. For this purpose the Contractor shall, subject to any provision to the contrary, provide all facilities that the SO may require. Where procedures are provided for or referred to in the Specification regarding specific tests the same shall be complied with. If at the discretion of the SO an independent expert is employed to make any such tests as are referred to in this paragraph, his charges shall be borne by the Contractor only if the tests disclose that such things are not in accordance with the provisions of the Contract. The report of the independent experts shall be final and conclusive.

Comment

1. Where test items are measurable under the Standard Method of Measurement, items should be provided in the bills of quantities against which the contractor is able to insert his price for testing. Alternatively provisional sums may be included for the cost of such tests.

2. If under test an item fails to comply with the specification in respect of workmanship or materials, the cost of testing will be borne by the contractor.

3. The contractor should never allow himself to be talked into a situation where he undertakes numerous tests without financial reimbursement on the pretext that Clause 13(3) entitles the SO to require him to carry out such tests.

Clause 13(4)—Workmanship to Comply with the Provisions of the Contract

The Works shall be executed in a workmanlike manner and to the satisfaction in all respects of the SO. If any things for incorporation do not accord with the provisions of the Contract or if any workmanship does not so accord the same shall at the cost of the Contractor be replaced, rectified or reconstructed as the case may be, and all such things which are rejected shall be removed from the Site.

Comment

1. The Contractor should always take care to ensure that if his work or materials are rejected that such rejection takes place because they do not conform with the requirements laid down in the various contract documents, namely the Specification, Bills of Quantities or Drawings and that rejection is not being made on the basis of the items concerned not conforming with the requirements of the SO where the SO's requirements are in fact not the same as those laid down in the Contract.

Clause 14—Local and other Authorities' Notices and Fees

The Contractor shall (so far as may be applicable to the Works) give all notices required by any Act of Parliament (whether general, local or personal) or by the regulations and/or bye-laws of any local authority and/or of any public service, company or authority affected by the Works or with whose systems the same are or will be connected, and he shall pay and indemnify the Authority against any fees or charges demandable by law under such Acts, regulations and/or bye-laws in respect of the Works and shall make and supply such drawings and plans required in connection with such notices.

Comment

1. In general, works carried out for the Crown are not affected by planning legislation and the normal service of notices will not be called for.

2. So far as the contractor may find himself liable to rate levies by the local authority on his temporary accommodation, he will have to meet

such charges himself without right to reimbursement as is normally the
case under both JCT and ICE Conditions.

Clause 15—Patent Rights

All royalties, licence fees or similar expenses in respect of the supply or use for or
in connection with the Works of any invention, process, drawing, model, plan or
information shall be deemed to have been included in the Contract Sum and the
Contractor shall indemnify the Authority from and against all claims and
proceedings, which may be made or brought against the Authority and any
damages, costs and expenses incurred by the Authority by reason of such supply
or use:
   Provided that where such supply or use has been necessary in order to comply
with any instructions given by the SO under the Contract, any royalty, licence fee
or similar expense payable by the Contractor in respect of such supply or use and
not provided for or reasonably contemplated by the Contract, shall be deemed
for the purposes of Condition 9(2) to be an expense properly and directly incurred
in complying with an SO's instruction other than one for an alteration in,
addition to, or omission from the Works.

Comment

The underlying reasoning of this Clause is extremely simple in that
should an expense arising from some royalty or patent right have been
incurred for something originally included in the contract, then the
contractor is liable for the matter without reimbursement; whereas if the
liability has arisen from an SO's instruction given subsequent to the
placing of the contract, the Contractor will be reimbursed.

STANDARD LETTERS

Reference: **GC/19**

To: The SO

Dear Sir

Request for the Reimbursement of Costs in Correcting Faulty Setting Out Attributable to Inaccurate Information Provided

We refer to drawing/drawings .... issued pursuant to Clause 12 of the conditions providing information to enable the Contractor to set out the works.

The works have been correctly set out in accordance with the information provided but it is now apparent that the information was inaccurate in that ....
....
and that as a result ....
....

Your instruction is therefore requested pursuant to Clause 7(1)(b) clarifying the identified discrepancy.

This discrepancy/[and] your required instruction we consider has caused/will cause us properly and directly to incur expense beyond that provided for in or reasonably contemplated by the contract and as soon as reasonably practicable we provide notice pursuant to Clause 9(1)(b)(ii) of the conditions requiring the reimbursement of such additional expense in accordance with Clause 9(2)(a).

Details of the costs incurred are appended/will be forwarded when available in substantiation of our application to be paid such sums by way of advance under Clause 40(5) of the Conditions.

Yours faithfully

Contractor Limited

Reference: **GC/20**

To: The SO

Dear Sir

Application for the Cost of Tests

We have recently supplied samples of ....
and carried out tests thereon in accordance with the requirements of the
SO pursuant to Clause 13(3) of the conditions.

Since such provision and testing were not clearly intended by or provided
for in the contract in accordance with the requirements of the Standard
Method of Measurement of .... and such
tests have shown the work/[and] the materials to comply in all respects
with the provisions of the contract, application is made for reimbursement of the costs concerned.

Supporting details are appended.

Yours faithfully

Contractor Limited

Reference: **GC/SO/16**

To: The Contractor

Dear Sir

Request to Rectify Setting Out Error

We have noted that .... is
wrongly set out [in relation to the setting out information provided in the
form of ....] and that this has arisen

through incorrect information supplied in writing to you in the form of
..../issued pursuant to
Clause 12.

You are therefore required to rectify the error at your own cost/and the
cost of so doing will be borne by the Authority pursuant to Clause
9(2)(a)(i) of the conditions.

[Details of your additional costs should be promptly submitted in
accordance with the requirements of the contract.]

Yours faithfully

Superintending Officer

Reference: **GC/SO/17**

To: The Contractor

Dear Sir

Instruction to Carry Out Test

You are directed, pursuant to Clause 13(3) of the conditions, to carry out
a test on .... at the place of
manufacture/fabrication/site. The actual material/article for testing will
be selected by Mr ....

The form of the test will be as set out in section .... of the
specification/as detailed on the appended schedule/as instructed by Mr
....

[As such test was clearly provided for in the Contract/designed to
ascertain whether the design of the finished work is appropriate for the
purposes intended and was particularised in the specification/bill of
quantities, the cost of making such test will be borne by yourself pursuant
to Clause 13(3) of the conditions.]

[Since such test was not provided for/particularised in the contract, the cost will be borne by the employer providing the test shows the workmanship or materials to be in accordance with the provisions of the contract and the SO's instructions.]

Yours faithfully

Superintending Officer

Reference: **GC/SO/18**

To: The Contractor

Dear Sir

Order for the Removal of Materials not in Accordance with the Contract

You are hereby instructed to remove the undernoted materials from the site which in the opinion of the SO are not in accordance with the Contract:

(a) ....

(b) ....

(c) ....

and to replace the said materials with proper and suitable materials.

This instruction is given pursuant to Clauses 7(1)(c) and 13(4) of the Conditions.

Yours faithfully

Superintending Officer

Reference: **GC/SO/19**

To: The Contractor

Dear Sir

Order for the Re-execution of Work not in Accordance with the Contract

You are hereby instructed to remove and properly re-execute the undernoted work which in the opinion of the SO is not in accordance with the Contract in respect of materials/[and] workmanship:

(a) ....

(b) ....

(c) ....

This instruction is given pursuant to Clauses 7(1)(d) and 13(4) of the Conditions.

[Any default on your part in delaying the proper re-execution of the said work will cause the Authority to have the work carried out by other means, rendering you liable for all expenses consequent thereon or incidental thereto [which will be deducted from monies due or which may become due to you] pursuant to Clause 8 of the conditions.]

Yours faithfully

Superintending Officer

Reference: **GC/SO/20**

To: The Contractor

Dear Sir

Instruction Issued to Ensure Compliance with Regulations

GCW1-F

It has been noted that the work as detailed in the drawings and specifications/our previous instruction reference .... is not in compliance with .... requiring ....

You are therefore instructed to .... to ensure compliance pursuant to Clause 14 Conditions and such instruction shall be deemed to be a variation order under Clause 9(2).

Yours faithfully

Superintending Officer

Reference: **GC/SO/21**

To: The Contractor

Dear Sir

Reimbursement of Fees Paid in Respect of Royalties and Patent Rights

We refer to your request for reimbursement dated .... reference .... together with sup- porting evidence in respect of licence fees/royalties/ .... paid by you in respect of .... to comply with SO Instruction .... in relation to the execution of the works/whose property or rights are or may be affected by the works.

It is acknowledged that such licence fees/royalties/ .... are a cost to be reimbursed pursuant to Clause 15 of the Conditions, since such cost arises from a variation and the appropriate amount will be included in our next certificate issued under Clause 42.

Yours faithfully

Superintending Officer

OUR REFERENCE: 11O453

Surveyors Bookshop

Norden House, Basing View, Basingstoke, Hants. RG21 2HN Tel: (0256) 55234 Fax: (0256) 841151

PLEASE FIND ENCLOSED BALANCE/~~PART~~ OF YOUR ORDER REFERENCE.

With compliments

Claims for non-delivery and shortages must be advised within 30 days from the date of despatch for UK orders, 90 days for overseas orders.

CHAPTER 9

The Clerk of Works, Sundry Liabilities, Excavated Materials, Antiquities and Fossils

Clause 16—Appointment of Resident Engineer or Clerk of Works

The Authority may appoint a Resident Engineer or a Clerk of Works and the Contractor shall admit him and his assistants to the Site. The Resident Engineer or Clerk of Works may exercise all the powers of the SO under Condition 13(1) and (2) and such other powers of the SO under the Contract as the SO may give notice of to the Contractor. The exercise of or failure to exercise such powers by the Resident Engineer or the Clerk of Works shall in no way limit or vary the ability of the SO to exercise such powers subsequently.

Comment

1. It is somewhat amusing in the context of this clause to see the conditions using the terminology of the government form of contract, the ICE Conditions and the JCT form all in the same breath.

2. Unless the contractor has been given specific notice to the contrary, the powers of the resident engineer or the clerk of works under these conditions are extremely limited, being restricted purely to ensuring compliance with the terms of the specification and drawings under Clauses 13(1) and (2). It is to be recalled that any notice given by the SO must be in writing to accord with the requirements of Clause 1(6). This is therefore as close as the conditions get to requiring delegated written notices of authority to be given to the contractor.

3. The powers under Clauses 13(1) and (2) in fact only relate to inspection and ensuring that the materials used for incorporation in the Works conform to the contract terms. In theory such right of inspection would not appear to extend to the workmanship, since (possibly inadvertently) that is covered under sub-section (4) of the clause, which is not among the areas of automatic authority bestowed.

4. No specific powers are given to the clerk of works or to the resident engineer to issue instructions under Clause 7 and he can only do so providing specific notice has been given to the contractor. In view of this, rather as under the JCT form, directions received from the clerk of works should be treated with considerable circumspection and written confirmation of them sought from the SO prior to any action being taken on them.

Clause 17—Watching, Lighting and Protection of the Works

The Contractor shall provide all watchmen necessary for the protection of the Site, the Works, and of all things (whether or not for incorporation) on the Site, during the progress of the execution of the Works, and shall be solely responsible for and shall take all reasonable and proper steps for protecting, securing, lighting and watching all places on or about the Works and the Site which may be dangerous to his workpeople or any other person.

Comment

1. This requirement calling on the contractor to watch and light the works occurs in all of the standard forms of contract. The liability of the contractor is summed up in the words 'take all reasonable and proper steps'. It is for such 'steps' that the contractor is solely responsible and which will be deemed to be covered by the price or rates given for the carrying out of the works.

2. Should the SO give instructions under Clause 7(1)(m) requiring the contractor to provide facilities or take precautions beyond such as could ever reasonably have been foreseen by an experienced contractor, the matter should be pursued as an item of claim under Clause 9(2).

Clause 18—Precautions to Prevent Nuisance

The Contractor shall take all reasonable precautions to prevent a nuisance or inconvenience to the owners, tenants or occupiers of other properties and to the public generally and to secure the efficient protection of all streams and waterways against pollution.

Comment

1. This type of requirement occurs again in all the standard forms of contract even though it may not always extend to specific mention of 'protection of all streams and waterways against pollution'.

2. The comments that were made in relation to Clause 17 concerning the SO giving instructions about such matters beyond such as could ever have been contemplated by the Contractor at time of tender apply equally to this requirement. Should such occur, the contractor should pursue the matter as a subject of claim under Clause 9(2).

Clause 19—Removal of Rubbish

The Contractor shall at all times keep the Site free from all rubbish and debris arising from the execution of the Works.

Comment

It seems a peculiarity of most standard forms of contract to become concerned with the cleanliness of the Site. In the writer's opinion this is a matter to be covered in the specification and it has always remained a mystery as to why it needs to feature in the Conditions as such.

Clause 20(1)—Materials Arising from Excavations

Subject to the provisions of paragraph (2) of this Condition, material of any kind obtained from the excavations shall remain the property of the Authority. Such material shall be dealt with as provided in the Contract, but the SO shall have power to direct its use in the Works or disposal by other means. When the Authority's property is permitted to be used in substitution for any things (whether or not for incorporation) which the Contractor would otherwise have provided the Quantity Surveyor shall ascertain the amount of any saving in the cost of the execution of the Works and the Contract Sum shall be reduced by the amount of any such saving.

Comment

1. There appears to be within this Clause an inherent source of difficulty. The disposal of excavated material will either have been described in the contract bill of quantities as 'cart away to tip provided by the Contractor', or 'cart away to tip provided by the employer giving the distance of haul', or 'return fill and ram the excavated material around foundations', or in some other way the mode of disposal will have been dealt with.

Difficulty may obviously therefore arise in those circumstances where in the original bills no indication has been given that the materials arising from excavation are to be incorporated in the works and where the contractor, realising that the excavated material will have some residual value, has taken it into account in his pricing, as might be done in the case of a cart-away item where the material could apparently be used and sold off the site as a filling material. The clause however states, 'material of any kind obtained from the excavation shall remain the property of the Authority', and therein lies a potential difficulty. The only answer would appear to be that the contractor should take no account of any residual value that excavated material might have in the preparation of his tender, since subsequently the Authority might elect to retain ownership in the material concerned.

All very well in theory but not in the context of a harsh competitive market where the contractor is looking for every means of saving money in his bid.

2. The only tenuous remedy open to the contractor would appear to be to obtain an instruction from the SO under Clause 7(1)(l) and to claim for the additional expense so caused under Clause 9(2). Even there it might be argued that the contractor has not incurred an additional expense but merely lost a source of additional income.

True it amounts to exactly the same thing in money terms to the contractor, but for the most part the contractor will not be dealing with people who think in a strictly commercial way.

On balance the matter had better be qualified at time of tender, since to leave the matter open would be a recipe for disaster. Imagine a situation in which the contractor was carting away from the site thousands of cubic metres of topsoil for which he had a market and had reflected that in his price, only be informed by the Authority that they would retain the topsoil!

Clauses 20(2) and (3)—Fossils and Antiquities

(2) All fossils, antiquities and other objects of interest or value which may be found on the Site or in carrying out excavations in the execution of the Works shall (so far as may be) remain or become the property of the Authority and upon discovery of such an object the Contractor shall forthwith:
(a) use his best endeavours not to disturb the object;
(b) cease work if and in so far as the continuance of work would endanger the object or prevent or impede its excavation or removal;
(c) takes steps which may be necessary to preserve the object in the exact position and condition in which it was found; and,
(d) inform the SO of the discovery and precise location of the object.
(3) The SO shall issue instructions in regard to what is to be done concerning an object the finding of which is reported to him by the Contractor under this Condition, which may include instructions requiring the Contractor to permit the examination, excavation or removal of the object by a third party.

Comment

1. The financial situation in which the contractor may find himself in the event of some antiquity or fossil being discovered is not very satisfactory. He must use his best endeavours not to disturb the object and the SO shall then issue instructions in regard to what is to be done concerning the object that has been found. Such instructions, probably not being an instruction involving variation or modification of the design, quality or quantity of the works or the addition or omission or substitution of any work, will fall for valuation within the net of the quasi-claim Clause 9(2) with all its ensuing likely payment delay, as previously discussed. Remember that in these conditions there is no equivalent to Clause 12 of the ICE Conditions.

2. The Contractor should therefore endeavour to argue that the matters involve an alteration, addition or omission to the contract works and should therefore be dealt with under Clause 9(1) and preferably on a daywork basis under sub-section (d).

STANDARD LETTERS

Reference: **GC/21**

To: The SO

Dear Sir

Request for Written Confirmation of an Instruction of the Resident Engineer/Clerk of Works

We have recently received a written/oral instruction from the Resident **Engineer/Clerk of Works Mr** .... on .... relating to .... and noting that the instruction does not fall within the authority of Clauses 13(1) or (2) or of any delegated authority notified to us pursuant to Clause 16 of the conditions, will be taking no action on it pending an SO's written confirmation.

Yours faithfully

Contractor Limited

Reference: **GC/22**

To: The SO

Dear Sir

Notification of the Discovery of Antiquities

We herewith give written notice of discovery of .... located at .... pursuant to Clause 20(2)(d) of the Conditions.

We are complying with the general obligations imposed on us in relation to fossils, antiquities and other objects of interest or value which may be found on the site under Clause 20(2) and request your instruction in regard to what is to be done concerning the object reported.

Yours faithfully

Contractor Limited

Reference: **GC/23**

To: The SO

Dear Sir

Notice of Additional Expense Arising from the Discovery of an Antiquity

We refer to our written notice of discovery of ....
.... located at
.... pursuant to Clause 20(2)(d) of the Conditions and to your subsequent instruction issued under Clause 20(3) and the general powers of instruction of Clause 7(1)(m).

That instruction directing that we .... is now considered to be causing us properly and directly to incur expense beyond that provided for in or reasonably contemplated by the contract, and as soon as reasonably practicable we provide notice pursuant to Clause 9(2)(b)(ii) of the Conditions requiring the reimbursement of the additional expense in accordance with Clause 9(2)(a).

Details of the costs incurred are appended/will be forwarded when available in substantiation of our application to be paid such sums by way of advance under Clause 40(5) of the Conditions.

Yours faithfully

Contractor Limited

Reference: **GC/SO/22**

To: The Contractor

Dear Sir

Notice of the Functions of an Assistant of the Superintending Officer

We hereby inform you that the undernoted person/persons is/are appointed as Assistant/Assistants of the Superintending Officer pursuant to Clause 16 of the Conditions.

[....]

[(a)..]

[(b)..]

[(c) ....]

His/Their functions/respective functions are as set out hereunder:-

[..

....]

[In the case of (a)...

....]

[In the case of (b)...

....]

[In the case of (c)...

....]

Any instruction given by him/them for those purposes shall be deemed to have been given by the Superintending Officer.

Yours faithfully

Superintending Officer

Reference: **GC/SO/23**

To: The Contractor

Dear Sir

Response to a Notice of Dissatisfaction with an Instruction of [an Assistant of] the Superintending Officer

We refer to your letter dated .... reference .... concerning an instruction of .... given concerning ....

We note the dissatisfaction expressed but confirm/and withdraw [and vary] the instruction/and now require that you should

....

....

We give this response pursuant to Clause 16 of the Conditions.

Yours faithfully

Superintending Officer

Reference: **GC/SO/24**

To: The Contractor

Dear Sir

Notification of Safety/[and] Security Requirement

We refer to the state of ....
.... and
consider the matter to be a safety/[and] security hazard.

You are therefore required to provide and maintain at your own cost lights/guards/fencing/warning signs adjacent to............................/ generally in connection with ....pursuant to Clause 17 of the Conditions.

Yours faithfully

Superintending Officer

Reference: **GC/SO/25**

To: The Contractor

Dear Sir

Notice of Nuisance/Inconvenience to Others

We refer to your construction operations [in the area of ....] and are concerned to note from our own observations/complaints received from members of the public/our client that the works are being carried out in a way that is interfering unnecessarily/improperly with the convenience of the public/access to ..../use by the public of ..../creating unreasonable noise/disturbance.

Your attention is drawn to your obligations and liabilities under Clause 18 of the Conditions and your potential liability for costs, charges and expenses arising out of the items that are the subject of complaint.

Yours faithfully

Superintending Officer

Reference: **GC/SO/26**

To: The Contractor

Dear Sir

Instruction Pertaining to Discovery of Antiquities

We refer to the recent discovery made of .... on
the site/in the area of ....
and give the undernoted instruction/instructions in relation to the
item/items pursuant to Clause 20(3) of the Conditions.

  (a) The ownership of the said item/items shall vest in the hands of the
      Authority.
  (b) All reasonable precautions shall be taken to prevent damage to the
      item/items.
  [( ) Facilities shall be provided to enable the items to be removed by
      ..../your own labour under the supervision of
      ....]
  [( ) The item/items shall be handed to ....]
  [( ) ....
      ....]

Records of any additional costs incurred shall be maintained for the SO's
approval and subsequent reimbursement by the Authority.

Yours faithfully

Superintending Officer

CHAPTER 10

Examination Prior to Covering up, Suspensions Due to Weather, Dayworks, Precautions and Damage Due to Accepted Risks

Clause 21—Examination of Excavations

The Contractor shall not lay any foundations until the excavations for the same have been examined and approved by the SO.

Comment

To avoid the cost consequences stemming from having the SO require foundations to be removed, it is essential that a notice requiring an inspection be regularly served on the SO prior to any concrete being poured in a way comparable to the requirements of Clause 22.

Clause 22—Notice Prior to Covering Up

The Contractor shall give reasonable notice to the SO whenever any work or thing for incorporation is intended to be covered in with earth or otherwise, and in default of doing so shall, if required by the SO, uncover such work and thing at his own expense.

Comment

1. Notice preparatory to covering up work will apply in many areas of a building and not only in the case of foundation works. Within the context of this clause the SO might well argue that prior to plastering a concrete wall he should have had the opportunity to inspect the keying that had been carried out before it was covered.
2. In order to avoid a flood of notifications on this subject, it will be advisable at the beginning of the contract to agree in writing with the SO those sections of the work that he specifically requires to inspect before a following trade covers the work concerned and thereafter to serve notice scrupulously against the agreed items.

Clause 23—Suspension for Inclement Weather

If the SO shall be of the opinion that the execution of the Works or any part thereof should be suspended to avoid risk of damage from frost, inclement

87

weather or other like causes, then, without prejudice to the responsibility of the Contractor to make good defective and/or damaged work, the SO shall have the power to instruct the Contractor to suspend the execution of the Works or any part thereof and the Contractor shall not resume work so suspended until permitted to do so by the SO. The Contractor shall not be entitled to any increase in the Contract Sum under Conditions 9(2) or 53(1) in respect of any expense incurred in consequence of any such instruction unless he can show that he has complied with all the requirements of the Specification relating to the avoidance of damage due to frost, inclement weather or other like causes.

Comment

1. This clause is one of the very few examples of the GC/Works/1 form being kinder to the contractor than most other forms.

2. If working conditions are becoming extremely difficult due to bad weather, it obviously behoves the contractor to try to persuade the SO that he should suspend the works in order to avoid damage.

3. Should the works have been so suspended, the Contractor shall be entitled to recover the additional expense incurred as a result of the suspension providing that he can show that he has complied with all the requirements of the specification designed to cover the avoidance of damage due to frost and inclement weather.

4. There have been a number of documents issued by the PSA over the years concerned with winter working. It is presumably in the light of the information and techniques highlighted in these documents that the current clause has been incorporated.

5. There would appear to be numerous instances where, regardless of the precautions that the contractor might be taking, an order for suspension to avoid damage would be appropriate. The ground formation in connection with roadworks readily springs to mind.

6. It would appear that the present set of Conditions have been drafted more with building than civil engineering in mind even though the title be 'General Conditions of Government Contracts for Building and Civil Engineering Works'.

Clause 24—Dayworks

The Contractor shall give to the SO reasonable notice of the commencement of any work ordered to be executed by daywork and shall deliver to the SO within one week of the end of each week vouchers in the form required by the SO giving full detailed accounts of labour, material and plant for that pay week. A copy of each voucher, if found correct, will be certified by the SO and returned to the Contractor.

Comment

1. At the beginning of the job the contractor should ascertain the type of daywork sheet that will be acceptable to the SO. Often the PSA will insist on its own standard daywork form being used and in that case the contractor has little choice but to acquiesce.

2. The clause further provides for certain conditions precedent being fulfilled by the contractor prior to him being able to obtain reimbursement for work on daywork; these conditions precedent are as follows:

    (a)   that notice of commencement of daywork shall have been given to the SO, and

    (b)   that no later than the end of the week following the week in which the work has been carried out, the daywork sheets shall have been submitted.

3. Unlike the ICE Conditions, the daywork arrangements under this contract do not appear to require the provision of an authenticated daily record sheet from which an actual daywork sheet is compiled but rather merely a daywork sheet itself submitted not later than the end of the week after that in which the work has been undertaken.

Clause 25(1)—Precautions Against Fire and Other Accepted Risks

The Contractor shall take all reasonable precautions to prevent loss or damage from any of the accepted risks and to minimise the amount of any such loss or damage or any loss or damage caused by a servant of the Crown. The Contractor shall comply with such instructions to this end as may given to him from time to time in writing by the SO.

Comment

1. The 'accepted risks' are those risks the consequences of which are accepted by the Authority and are defined under Clause 1(2) as being:

    (a)   fire or explosion,

    (b)   storm, lightning, tempest, flood or earthquake,

    (c)   aircraft or other aerial devices or objects dropped therefrom, including pressure waves caused by aircraft or such devices whether travelling at sonic or supersonic speeds,

    (d)   ionising radiations or contamination by radioactivity from any nuclear fuel or from any nuclear waste from the combustion of nuclear fuel,

    (e)   the radioactive, toxic, explosive or other hazardous properties of any explosive nuclear assembly or nuclear component thereof,

    (f)   riot, civil commotion, civil war, rebellion, revolution, insurrection, military or usurped power of King's enemy risks.'

2. It is only reasonable therefore that the contractor should be expected to take precautions to prevent an occurrence that would involve the Authority in considerable loss and that he should take instructions from the SO about such matters.

If, however, the nature of the precautions instructed by the SO were totally outside the reasonable contemplation of the contract and could certainly never have been foreseen, either specifically or by implication from the information contained in the contract documents, then the contractor might, with some justification, lodge a claim for the additional expense involved under Clause 9(2).

Clause 25(2)—Compliance with Regulations Governing the Storage of Explosives and Petrol.

The Contractor shall comply with any statutory regulations (whether or not binding on the Crown) which govern the storage of explosives, petrol, or other things (whether or not for incorporation) which are brought on the Site.

Comment

1. In the execution of construction works the Crown does not necessarily have to abide by the various byelaws and regulations that govern construction work in other areas of the public sector and that would apply to private clients.

2. Because of this, this requirement for compliance, notwithstanding such general exoneration, is called for in the case of explosives, petrol and presumably other like products of a dangerous nature.

Clause 26—Responsibility for Damage to Works

(1) All things not for incorporation which are on the Site and are provided by or on behalf of the Contractor for the construction of the Works shall stand at the risk and be in the sole charge of the Contractor, and the Contractor shall be responsible for, and with all possible speed make good, any loss or damage thereto arising from any cause whatsoever, including the accepted risks.

(2) (a) The Contractor shall (unless the Authority exercises his powers to determine the Contract) with all possible speed make good any loss or damage arising from any cause whatsoever occasioned to the Works or to any things for incorporation on the Site (including any things provided by the Authority) and shall notwithstanding any such loss or damage proceed with the execution and completion of the Works in accordance with the Contract.

(b) The cost of making good such loss or damage shall be wholly borne by the Contractor, save that:

(i) where the loss or damage is wholly caused by the neglect or default of a servant of the Crown acting in the course of his employment as such, the Authority shall pay the Contractor for making good the loss or damage, and where it is partly caused by such neglect or default, the Authority shall pay the Contractor such sum as is proportionate to that servant's share in the responsibility for the loss or damage, and,

(ii) where the loss or damage is wholly caused by any of the accepted risks the Authority shall pay the Contractor for making good the loss or damage and where it is partly so caused the Authority shall pay the Contractor such sum as is proportionate to the share of any of the accepted risks in causing the loss or damage.

(c) Any sum payable by the Authority under sub-paragraph (2)(b) of this Condition shall be ascertained in the same manner as a sum payable in respect of an alteration or addition under the Contract and shall be added to the Contract Sum.

Comment

1. In résumé, the cost liability for reinstatement in the event of damage rests between the contracting parties in the following way:

(a) Plant and temporary accommodation are solely at the contractor's risk, regardless of the cause or origin that gave rise to the damage.

(b) Loss or damage caused to the permanent works, or to materials on site for incorporation in the works, caused by the neglect or default of a servant of the Crown shall be a charge to the Authority.

(c) Loss or damage to the permanent works or to materials for incorporation therein due to the accepted risks shall be to the cost of the Authority and treated and valued as though it were a variation order.

2. The contractor shall make good all damage as soon as possible.

3. The valuation of repair work where the cost is to be borne by the Authority will be carried out in accordance with Clause 9 of the conditions and may give ground for an extension of time under Clause 28(3)(e).

Clause 27—Assignment or Transfer of Contract

The Contractor shall not, without the consent in writing of the Authority, assign or transfer the Contract, or any part, share or interest therein. No instalment or other sum of money to become payable under the Contract shall be payable to any other person than the Contractor unless the consent of the Authority in writing to the assignment or transfer of such money to such person be produced when such payment is claimed as due.

Comment

1. Assignment covers a situation in which one party is substituted for by another party and the former ceases to exist contractually. Thus the substitute party undertakes all of the obligations and liabilities of the former party.

2. It is a general requirement of most contracts that this arrangement can only be undertaken with the permission of the employing party. Such an arrangement is after all eminently reasonable, since the former party was the one chosen to undertake the service in the first place and it would be totally unreasonable for those obligations to be passed on willy-nilly to another party without the employer, or in this case as he is known the Authority, having acquiesced.

STANDARD LETTERS

Reference: **GC/24**

To: The SO

Dear Sir

Notification of the Availability of Work for Examination and/or Measurement Before Covering Up

We give notice pursuant to Clauses 21/[and]/22 of the conditions that .... is available for inspection prior to laying foundations/[and]/measurement before covering up.

We request that inspection/[and]/measurement be made by .... to enable the works to proceed.

Yours faithfully

Contractor Limited

Reference: **GC/25**

To: The SO

Dear Sir

Clarification of the Occasions on Which the SO will Require Notice Prior to Covering Up Work

We are aware, under the terms of Clause 22 of the conditions, that notice is required to be given prior to work being covered up.

Having regard to the nature of the works being undertaken, there are numerous cases where following trades will be covering up the work of preceding trades and some guidance from you as to the occasions on which notice will be required would be appreciated.

Yours faithfully

Contractor Limited

Reference: **GC/26**

To: The SO

Dear Sir

Notification of Additional Expense Arising from the Suspension of Work Due to Inclement Weather

We refer to your instruction dated .... issued pursuant to Clause 23 of the Conditions instructing that the execution of [part of] the works [being ....] be suspended to avoid risk of damage from frost/inclement weather/... .

Noting that we have complied in every respect with the terms of the specification relating to the avoidance of damage due to frost, inclement weather or other like causes, we now consider such instruction to be causing us properly and directly to incur expense beyond that provided for in or reasonably contemplated by the contract and as soon as reasonably practicable we provide notice pursuant to Clause 9(2)(b)(ii) of the conditions requiring the reimbursement of such additional expense in accordance with Clauses 9(2)(a) and 23.

Details of the costs incurred are appended/will be forwarded when available in substantiation of our application to be paid such sums by way of advance under Clause 40(5) of the Conditions.

Yours faithfully

Contractor Limited

Reference: **GC/27**

To: The SO

Dear Sir

Notification of Dayworks

We herewith give notice pursuant to Clause 24 of the Conditions that [daywork is to be commenced in accordance with your instruction dated .... relating to ....] [in our opinion .... cannot be properly measured and valued] and we therefore request 'daywork' payment for the work concerned pursuant to Clause 9(1)(d) of the conditions.

Records of the labour, plant and materials employed will be submitted for verification not later than the end of the week following that in which the work has been executed.

Yours faithfully

Contractor Limited

Reference: **GC/28**

To: The SO

Dear Sir

Submission of Daywork Records and/or Accounts

In accordance with the requirements of Clause 24 of the Conditions we append our [priced] record/schedules of daywork covered by our Notification No. .... dated .... relating to ....

....

Yours faithfully

Contractor Limited

Reference: **GC/29**

To: The SO

Dear Sir

Notice of Claim Arising from Unforeseen Precautions in Respect of Accepted Risks

We refer to your instruction dated .... issued pursuant to Clause 25(1) and the powers of instruction bestowed under Clause 7(1)(m) of the conditions relating to precautions to be taken to prevent loss or damage from.. ...., being one of the 'accepted risks' defined under the contract.

We consider such instruction to be causing us properly and directly to incur expense beyond that provided for in or reasonably contemplated by the contract and as soon as reasonably practicable we provide notice pursuant to Clause 9(2)(b)(ii) of the conditions requiring the reimbursement of such additional expense in accordance with Clause 9(2)(a).

Details of the costs incurred are appended/will be forwarded when available in substantiation of our application to be paid such sums by way of advance under Clause 40(5) of the Conditions.

Yours faithfully

Contractor Limited

Reference: **GC/30**

To: The Authority

Dear Sir

Application for Reimbursement of Reinstatement Costs

Loss and damage has recently been occasioned to the works [and things for incorporation therein on site] and notwithstanding this we are proceeding with the execution and completion of the works in accordance with our contract obligations.

However, noting such loss and damage to have been wholly/partially caused by [the neglect or default of a servant of the Crown acting in the course of his employment] [one of the 'accepted risks' defined under the contract] in that ....
we make application to be paid for the reinstatement of such loss and damage pursuant to Clause 26(2)(b) of the Conditions in whole/in proportion to ....
ascertained in accordance with Clause 26(2)(c).

Details of the sums involved are appended/will be forwarded when available in substantiation of our application to be paid such sums by way of advance under Clause 40(5) of the Conditions.

Yours faithfully

Contractor Limited

Reference: **GC/SO/27**

To: The Contractor

Dear Sir

Approval/Disapproval of Construction Excavation

It is confirmed that inspection of excavations located at:

(i) ....

(ii) ....

will be carried out by .... on
.... at ....</has been carried out by
.... and that approval for foundation
construction has been given/has been denied pursuant to Clause 21 of the
Conditions.

Yours faithfully

Superintending Officer

Reference: **GC/SO/28**

To: The Contractor

Dear Sir

Procedure to be Adopted Prior to Covering Up Work

Your attention is drawn to the fact that prior notice is required to be
given by you before any work is covered up or put out of view pursuant
to Clause 22 of the Conditions.

In this connection we require you to notify ....
to enable any necessary measurement or examination to be carried out
particularly in respect of but not limited to:

(a) ....

(b) ....

(c) ....

( ) ....

[In the event that your company already operates appropriate notification procedures that will meet our approval, we will be pleased to adopt your existing arrangements.]

[The required notifications will be given on our 'Availability for Inspection Form', a copy of which is appended for your information.]

Yours faithfully

Superintending Officer

Reference: **GC/SO/29**

To: The Contractor

Dear Sir

Availability for Inspection

Receipt of 'Availability for Inspection Form' reference number .... is acknowledged.

It is confirmed that inspection in this particular case is not considered necessary/will be carried out by .... on .... at ..../has already been carried out by .... and that approval for covering up will be signed for at time of inspection when passed/has been given/has been denied pursuant to Clause 22 of the Conditions.

Yours faithfully

Superintending Officer

Reference: **GC/SO/30**

To: The Contractor

Dear Sir

Instruction to Uncover Work

You are hereby instructed to uncover...
..../make   an   opening   through
.... pursuant to Clause
22 of the Conditions.

As you have [not] complied with the requirements of Clause 22 of the
Conditions prior to covering-up the said work, any cost in so doing and
subsequent reinstatement will be borne by yourself/the Authority
providing the works are found to be executed in accordance with the
terms of the Contract.

Yours faithfully

Superintending Officer

Reference: **GC/SO/31**

To: The Contractor

Dear Sir

Suspension of Work Due to Inclement Weather

We hereby instruct you to suspend the progress of the works/that part of
the Works being ....
pursuant to Clause 23 of the Conditions.

[As such suspension arises from a cause attributable to weather
conditions/attributable to your default in that ....
....,

this is not a matter giving entitlement to financial compensation [nor to extension of time] since the suspension is provided for in the contract/arises from your default.]

During the period of suspension you are required to protect and secure the Work properly [and in particular you are to ....].

[It is acknowledged that such suspension provides ground [both] for financial compensation/[and] extension of time and you are strongly advised to comply with the various provisions of the Contract in regard thereto.]

Yours faithfully

Superintending Officer

Reference: **GC/SO/32**

To: The Contractor

Dear Sir

Lifting of Suspension of Work

We refer to our notification of suspension of work dated .... reference .... and now confirm its lifting as from .... pursuant to Clause 23 of the Conditions.

Yours faithfully

Superintending Officer

Reference: **GC/SO/33**

To: The Contractor

Dear Sir

Daily Record of Resources Used on Daywork

The 'Daily Record of Resources Used on Daywork' in respect of
.... dated
.... is acknowledged.

The record is agreed correct and has been signed/has been amended in
consultation with your representative and signed as amended/is rejected
as the work covered by the record is already included elsewhere under the
Contract and we draw you attention to..
..../no instruction for the work was given
[although] it has been rendered out of time.

One copy is returned to you pursuant to Clause 24 of the Conditions.

Yours faithfully

Superintending Officer

Reference: **GC/SO/34**

To: The Contractor

Dear Sir

Daywork Sheets

Receipt of your daywork sheet/sheets number ..../numbers
.... to .... for the week of ....
is/are acknowledged.

[The items agree with your previously submitted records and the pricing
is in accordance with the Contract and arithmetically correct/and
although certain arithmetical mistakes were made, these have been
corrected. The sheet/sheets has/have been signed and one copy is
returned to you for inclusion in the accounts.]

[The items are unsupported by agreed records/not priced in accordance with the contract/submitted out of time and are therefore rejected.]

These procedures are operated pursuant to Clause 24 of the Conditions.

Yours faithfully

Superintending Officer

Reference: **GC/SO/35**

To: The Contractor

Dear Sir

Notification to Comply with Precautions against Loss/Damage/...............

We refer to the state of ....
and consider that loss/damage/.................... may result.

You are therefore required to take all reasonable precautions to provide and maintain adequate services to minimise any loss/damage/..............
in connecton with... pursuant to Clause 25(1) of the Conditions.

Yours faithfully

Superintending Officer

CHAPTER 11

Date for Completion or Part Completion, Extensions of Time and Liquidated Damages

Clause 28(1)—Date for Completion

The Works shall be carried on and completed to the satisfaction of the SO and all unused things for incorporation and all things not for incorporation, removal of which is ordered by the SO, shall be removed and the Site and the Works cleared of rubbish and delivered up to his satisfaction on or before the date for completion.

Comment

1. The date for completion will be the date named in the contract documents or any later date for which a contract extension has been granted.

2. As in most other standard forms of contract, the contractor's obligation is to complete the works 'on or before the date for completion'. Thus the contractor has a choice when programming the work either to finish on the date for completion or at some time prior to the date for completion. Providing such earlier completion is mutually accepted by the parties to the contract as being achievable and the actions of the parties provide a clear demonstration of the intention to achieve such earlier completion, it is to be contemplated that the contractor could be delayed in achieving such earlier completion, and if the delay was due to a default of the Authority, look to the Authority for compensation.

3. Earlier versions of the government form of contract used the words 'perfect state' in relation to the state of completion at the completion date under the contract. The conditions now merely require the work to be 'completed to the satisfaction of the SO' and in practice this now means that the state of completion will be that normally employed in most standard conditions of contract—i.e. a state of practical or substantial completion.

4. Under other forms of contract such state of completion is generally interpreted to mean the works being in a sufficient state of completion to enable them to be used for the purposes for which they were intended and having passed any test prescribed under the contract but without necessarily being complete in every respect.

Clause 28(2)—Extension of Time

The Contractor shall be allowed by the Authority a reasonable extension of time for the completion of the Works in respect of any delay in such completion which has been caused or which the Authority is satisfied will be caused by any of the following circumstances:

(a) the execution of any modified or additional work;
(b) weather conditions which make continuance of work impracticable;
(c) any act or default of the Authority;
(d) strikes or lock-outs of workpeople employed in any of the building, civil engineering or analogous trades in the district in which the Works are being executed or employed elsewhere in the preparation or manufacture of things for incorporation;
(e) any of the accepted risks; or
(f) any other circumstance which is wholly beyond the control of the Contractor;

Provided that:

(i) except in so far as the Authority shall otherwise decide, it shall be a condition upon the observance of which the Contractor's right to any such extension of time shall depend that the Contractor shall, immediately upon becoming aware that any such delay has been or will be caused, give notice to the SO specifying therein the circumstances causing or likely to cause the delay and the actual or estimated extent of the delay caused or likely to be caused thereby;

(ii) the Contractor shall not be entitled to any extension of time in respect of a delay caused by any circumstance mentioned in sub-paragraph (2)(f) of this Condition if he could reasonably be expected to have foreseen at the date of the Contract that a delay caused by that circumstance would, or was likely to, occur;

(iii) in determining what extension of time the Contractor is entitled to the Authority shall be entitled to take into account the effect of any authorised omission from the Works;

(iv) it shall be the duty of the Contractor at all times to use his best endeavours to prevent any delay being caused by any of the above-mentioned circumstances and to minimise any such delay as may be caused thereby and to do all that may reasonably be required, to the satisfaction of the SO, to proceed with the Works; and

(v) the Contractor shall not be entitled to an extension of time if any such delay is attributable to any negligence, default or improper conduct on his part.

Comment

1. It should be appreciated that this clause deals solely with time and not with the contractor's right to compensation in respect of additional time. In general if the ground for extension of time is neither the fault of the contractor nor the Authority, for example, weather conditions, the contractor may have entitlement to an extension of time, but solely to the

extent of exonerating him from liquidated damages. If, however, the need
for the extension of time arises because of a default of the Authority, the
Contractor would have entitlement both to prolongation and to
disruption expenses under Clause 53.

2. These Conditions lack the sophistication of both JCT 80 and the
ICE Conditions as regards extensions of time in that there is no specific
requirement on the part of the Authority or SO to grant interim
extensions of time and to keep the contractor properly informed by
notification as to whether he is or is not entitled to extension. Clearly it is
good practice for the SO to do so but it is not a stated contractual
requirement. This tends to encourage an attitude to extensions of time on
the part of the SO of 'we will sort it all out at the end'—meanwhile the
contractor is running blind as regards the completion date.

3. The stated grounds for extension are now considered in turn.

(a) 'The execution of any modified or additional work.'
Where the execution of modified or additional work leads to an increase
in the value of work that is less than proportional to a pro-rata increase
in time, it is likely that the Contractor will have ground to claim for
additional preliminaries and overhead cover.

Furthermore, where the preliminaries and overhead cover have been
totally priced in the preliminaries bill and not reflected in the unit rates,
the under-recovery of time-related costs will become more marked and in
the latter case it is likely that the whole cost will be directly time related.

The execution of modified or additional work can only have taken
place as a result of an SO's instruction issued under Clause 7(1)(a) and as
such would provide a basis for compensation of prolongation and
disruption expenses under Clause 53(1)(a), subject of course to the
contractor having complied with the various procedural requirements of
the conditions as a condition precedent to recovery.

(b) 'Weather conditions which make continuance of work
impracticable.'
An extension of time granted under this head is unlikely to enable
financial recovery to be made unless it can be achieved within the
framework of the previously noted Clause 23. For this to happen, it
would be necessary for the SO to have issued an instruction to the
contractor requiring him to suspend the execution of the works, since to
continue operations would cause damage to the work, despite the fact
that the contractor had complied with all the requirements of the
specification relating to the avoidance of damage due to inclement
weather.

Unless it is quite clear that the terms of Clause 23 can be implemented,
the contractor should beware of being granted extensions of time on
the ground of bad weather when in reality they were merely covering

some deficiency of the Authority that would have entitled the contractor to both extra time and extra money.

(c) 'Any act or default of the Authority.'

An extension granted under this sub-clause will be the one most likely to give the contractor grounds for financial recovery against matters that are in all probability a common law breach of contract. The grounds for financial recovery are likely to exist under Clause 53(1)(d), read in conjunction with the matters listed under Clause 53(2).

The listed items of Clause 53(2) cover delay in the provision of design information, in the supply of materials or in the execution of work by the Authority and late nomination.

These three categories of default are among the most frequent common law breaches, although they by no means cover all the breaches that can occur. Where matters occur outside the coverage of Clause 53(2), it may very well be necessary to claim under common law entitlement.

(d) 'Strikes or lock-outs of workpeople employed in any of the building, civil engineering or analogous trades in the district in which the Works are being executed or employed elsewhere in the preparation or manufacture of things for incorporation.'

The strike or lock-out ground for extension will not normally provide a basis for financial recompense. Further, it is not clear from the form of words used in the clause as to whether strikes in the transport services engaged in the movement of materials about the country would also rate for an extension of time.

(e) 'Any of the accepted risks.'

It will be recalled that the accepted risks as defined in Clause 1(2) are those risks the consequences of which are accepted by the Authority. The most significant of them from a practical working point of view is fire. Under Clause 25(1) the contractor is required to take all reasonable precautions to prevent loss or damage from the accepted risks and in this respect to comply with the instructions of the SO. If, however, despite precautions having been taken, loss or damage is still caused, the Authority pays the contractor to make good the loss or damage pursuant to Clause 26(2)(b)(ii), with extension of time if necessary being granted under the clause we are now considering. Any time-related costs involved fall within the obligation of the Authority to pay the contractor.

(f) 'Any other circumstance which is wholly beyond the control of the Contractor.'

We have here the equivalent of the 'force majeure' provision of the JCT forms of contract and the 'any other special circumstances whatsoever' of the ICE Conditions.

The use of the words 'wholly beyond' in the sub-clause would appear to make it extremely difficult to obtain an extension of time under this heading. Even if one were obtained, it is most unlikely that it would arise from a circumstance providing entitlement to financial recompense.

4. As in all of the standard forms of contract, the service of the relevant notice to obtain an extension of time is both essential and legally a condition precedent. The matter is specifically covered under sub-section (2)(i) of the clause, requiring:

(i) Notice to be given immediately upon becoming aware of a delay. The word 'immediately' obviously has an extremely onerous impact and allows the contractor no time to cogitate as to whether or not he should serve notice. The golden rule on a GC/Works/1 contract must be to serve notice immediately and worry about the consequences afterwards.

(ii) The reason for the delay must be stipulated.

(iii) An estimate of the time involved is to be given.

5. The right of the contractor to receive an extension of time is subject to an important and overriding proviso in that the contractor shall not be entitled to an extension of time arising from a cause that he could reasonably have been expected to foresee at the date of tender.

However, under no circumstance should a contractor ever be prepared to concede the non-granting of an extension of time under this provision on the ground that he should have appreciated that the information he was going to receive would be inadequate and that he should have based this conclusion on the apparent adequacy or inadequacy of the documentation available at time of tender.

It must be emphasised that it is no part of a contractor's obligation to make a subjective judgement as to the adequacy of the manner in which the Authority's obligations under the contract will be fulfilled. In competitive tendering, the contractor, and all other contractors tendering, can only assume that those obligations will be adequately fulfilled and were individual contractors to reason otherwise, it could only make meaningless any form of competitive tendering.

6. When determining an extension of time the Authority is entitled to take into account the counteracting effect of any authorised omission from the works. In considering the effect of an omission, regard must be had to the time within the contract period at which the omission is actually ordered. It may very well be that an omission ordered at a late date on a contract that has already been the subject of considerable variation, could not be assimilated into the programme in a sufficiently meaningful way to reflect the apparent time saving that ought to have been made, had the knowledge of the omission been given earlier.

7. It is both stated in the clause and a commom law obligation of the contractor that he should at all times use his best endeavours to mitigate any delay that was caused by any of the specified reasons. On occasion, efforts on a contractor's part to fulfil this obligation will tend merely to exacerbate the damage that he is suffering if the works are in a continuing delay and disruption situation, with one cause of delay merely being followed by another.

Beware the situation in which good resources are being progressively thrown down the drain in a frantic endeavour to achieve the impossible against a continuing flood of inadequate performances and breaches on the part of the Authority.

8. Always appreciate that the contractor will not be entitled to an extension of time where the delay is attributable to his own negligence.

Clause 28A(1)—Sectional Completion

The Authority may, before the completion of the Works, take possession of any part of the Works (in this Condition referred to as a 'completed part') which is certified by the SO as having been completed to his satisfaction and is either:

(a)   a section specified in the Abstract of Particulars; or
(b)   a part of the Works (including a part of a section) in respect of which the parties agree, or the SO has given an instruction, that possession shall be given before completion of the Works;

and such completed part, on and after the date on which the certificate is given, shall no longer be deemed to form part of the Works for the purposes of Conditions 3 and 26.

Comment

1. This clause recognises the common situation in which the Authority takes possession of part of the works prior to completion. Such part may have been specified in the abstract of particulars, or be the subject of an SO's instruction, or have been the subject of mutual agreement.

2. The maintenance period for the part starts on completion of the part and as soon as possible after its completion the SO will certify its value for the purpose of calculating a reduction in any potential liquidated damages. Consequently should liquidated damages need to be levied, the amount levied will be reduced in proportion to the value of that part of the works previously handed over.

3. Of greater practical significance in the run-of-the-mill contract will be the fact that the retention fund will be reduced by a proportion in a similar manner, while the Contractor will no longer be liable for the care of the Works so far as it relates to the section.

Clause 29—Liquidated Damages

(1) If the Works shall not be completed and the Site cleared on or before the date for completion, the Contractor shall pay to the Authority liquidated damages in

respect of the delay in completion calculated in accordance with the details set out in the Abstract of Particulars for the period during which the Works shall remain uncompleted and the Site not cleared after the date for completion.

(2) No payment or concession to the Contractor or order for modified or additional work at any time given to the Contractor or other act or omission of the Authority or his servants shall in any way affect the rights of the Authority to recover the said liquidated damages or shall be deemed to be a waiver of the right of the Authority to recover such damages unless such waiver has been expressly stated in writing signed on behalf of the Authority.

(3) If at any time the Authority (whether or not he has previously allowed the Contractor any extension of time under Condition 28) gives notice to the Contractor that, in the opinion of the Authority, the Contractor is not entitled to any or (as the case may be) any further extension, then any sum which at that time would represent the amount of liquidated damages payable by the Contractor under this Condition or this Condition as modified by Condition 28A(4) (as the case may be) shall be treated for the purposes of Condition 43 as a sum recoverable from or payable by the Contractor.

Comment

1. In the event of the works not being completed by the date of completion named in the contract, or by any later date arising as a result of extension, the Contractor is liable to pay the Authority liquidated damages at the rate set out in the contract.

2. Recovery of liquidated damages may be made by the Authority against the individual contract concerned or, having regard to Clause 43, against any other contract with the Authority or with any department or office of Her Majesty's Government.

3. Prior to the Authority being entitled to recover liquidated damages, notice must have been given by the Authority to the contractor stipulating that he is not entitled to extension of time or any further extension of time. The giving of this notice is a condition precedent to the Authority's right of recovery.

It will be recalled, having regard to Clause 1(6), that the notice must have been given in writing.

STANDARD LETTERS

Reference: **GC/31**

To: The SO

Dear Sir

Application for Extension of Time

We give written notice pursuant to Clause 28(2)(i) of the Conditions of our having become aware of a delay that has been/will be caused to the works and that the completion of the works is likely to be/has been delayed in relation to the date on or before which the works were due for completion [as previously extended].

The circumstances giving rise to delay/This most recent circumstance giving rise to delay is .... Further we believe that we could not reasonably have been expected to foresee at the date of the contract that a delay caused by that circumstance would be likely to, or would occur.

This delay [and delays previously notified] with its/their associated request/requests for a reasonable extension to be allowed by the Authority are registered under the stipulated clause of the conditions on the ground of:

Clause		Reason	Time in Weeks
28(2)(a)	[(*)	varied works]	
	[(*)	work carried out against provisional sums]	
	[(*)	additional work]	
28(2)(b)	[( )	frost]	
	[( )	inclement weather]	
	[(*)	postponement of work due to inclement weather under Clause 23]	
28(2)(c)	[(*)	late receipt of setting out information]	

28(2)(c) cont.	[(*)	late receipt of drawings]
	[(*)	late receipt of details]
	[(*)	late receipt of bending schedules]
	[(*)	late nomination of a nominated sub-contractor]
	[(*)	late nomination of a nominated supplier]
	[(*)	late instruction]
	[(*)	non-availability of the site]
	[(*)	non-availability of part of the site]
	[(*)	late provision of free issue materials]
	[(*)	late completion of work by other contractors in the direct employment of the Authority]
28(2)(d)	[( )	strike or lock-out at the site]
	[( )	strike or lock-out at the works of a supplier]
28(2)(e)	[(*)	unforeseen precautions in respect of accepted risks under Clause 25(1)]
	[(*)	reinstatement of damage arising from accepted risks under Clause 26(2)(b)]
28(2)(f)	[( )	force majeure]
	[( )	strike or lock-out of transport workers affecting material deliveries]
	[(*)	clarification of ambiguities under Clause 7(1)(b)]
	[(*)	order of execution of work under Clause 7(1)(e)]
	[(*)	limitation of working hours under Clause 7(1)(f)]
	[(*)	suspension of work under Clause 7(1)(g)]
	[(*)	Opening up for inspection

		under Clause 7(1)(i)]
28(2)(f)	[(*)	Execution of emergency work
cont.		under Clause 7(1)(k)]
	[(*)	faulty setting out based on wrong information supplied under Clause 12]
	[(*)	discovery of antiquities under Clause 20(3)]
	[(*)	delay by an artist in the direct employ of the Authority under Clause 50]
	[(*)	delay by an independent contractor under Clause 50]
	[(*)	delay by a statutory authority under Clause 50]

We deem all of these weeks/..................... of these weeks to rate for re-imbursable loss and expense pursuant to Clause/Clauses 9(2)/20(3)/23/26(2)(b)/53 of the conditions, the appropriate notice/notices having been given.

We further undertake to use our best endeavours to prevent any delay being caused by any of the above-mentioned circumstances and to minimise any such delay as may be caused thereby and to do all that may reasonably be required, to the satisfaction of the SO, to proceed with the works.

Yours faithfully

Contractor Limited

Comment

All grounds for extension marked with an asterisk will probably carry ground for financial reimbursement. In some cases where a ground for extension has not been marked with an asterisk but where delay is caused under that heading as a direct consequence of an earlier delay under an asterisked heading, money will still be reimbursable. (For example, delay in the issue of instructions might result in the works falling back into a winter period giving rise to exceptionally adverse weather conditions which would not have been encountered had the earlier delay not have taken place).

Reference: **GC/SO/36**

To: The Contractor

Dear Sir

Extension of Time for the Works/a Section of the Works During the Contract Period/at the Due Date/or the Extended Date for Completion

We refer to your application/applications dated.................................
reference .... made for extension of time to the Works/that section of the Works being .... to which you consider you are entitled pursuant to Clause 28(2)(i) of the Conditions.

We have carefully investigated such claim/claims and consider that you are entitled to a total/cumulative extension to date in respect of the Works/ the noted section of the Works amounting to .... days..

The said total/cumulative extension is awarded pursuant to the under-noted stipulated clause or clauses of the Conditions on the ground of:

Clause		Reason	Time in Weeks
28(2)(a)	[(*)	varied works]	
	[(*)	work carried out against provisional sums]	
	[(*)	additional work]	
28(2)(b)	[( )	frost]	
	[( )	inclement weather]	
	[(*)	postponement of work due to inclement weather under Clause 23]	
28(2)(c)	[(*)	late receipt of setting out information]	
	[(*)	late receipt of drawings]	
	[(*)	late receipt of details]	
	[(*)	late receipt of bending schedules]	
	[(*)	late nomination of a nominated sub-contractor]	

28(2)(c) cont.	[(*)	late nomination of a nominated supplier]
	[(*)	late instruction]
	[(*)	non-availability of the site]
	[(*)	non-availability of part of the site]
	[(*)	Late provision of free issue materials]
	[(*)	Late completion of work by other contractors in the direct employment of the Authority]
28(2)(d)	[( )	strike or lock-out at the site]
	[( )	strike or lock-out at the works of a supplier]
28(2)(e)	[(*)	unforeseen precautions in respect of accepted risks under Clause 25(1)]
	[(*)	reinstatement of damage arising from accepted risks under Clause 26(2)(b)]
28(2)(f)	[( )	force majeure]
	[( )	strike or lock-out of transport workers affecting material deliveries]
	[(*)	clarification of ambiguities under Clause 7(1)(b)]
	[(*)	order of execution of work under Clause 7(1)(e)]
	[(*)	limitation of working hours under Clause 7(1)(f)]
	[(*)	suspension of work under Clause 7(1)(g)]
	[(*)	opening up for inspection under Clause 7(1)(i)]
	[(*)	execution of emergency work under Clause 7(1)(k)]
	[(*)	faulty setting out based on wrong information supplied under Clause 12]
	[(*)	discovery of antiquities

28(2)(f)   [(*)   under Clause 20(3)]
cont.   Delay by an artist
                          in the direct employ
                          of the Authority under
                          Clause 50]
                  [(*)   Delay by an independent
                          contractor under
                          Clause 50]
                  [(*)   Delay by a statutory
                          authority under
                          Clause 50]

                          Total of weeks granted

Yours faithfully

Superintending Officer

Note
All grounds for extension marked with an asterisk will probably carry grounds for
financial reimbursement. In some cases where a ground for extension has not been
marked with an asterisk but where delay is caused under that heading as a direct
consequence of an earlier delay under an asterisked heading, money may still be
reimbursable (for example, delay in the issue of an instruction might result in work
falling into the winter period or into a period of exceptionally inclement weather
which would not have been encountered had the earlier delay not taken place).

Reference: **GC/SO/37**

To: The Contractor

Dear Sir

*Notification that the Contractor is not Entitled to [Further] Extension of
Time During the Contract Period/at the Due Date or Extended Date of
Completion*

We have carefully investigated your application/applications for
extension of time for [all the circumstances affecting] the works/that
section of the works being..

We now confirm that we are unable to grant at this time/at this, the due date for completion [as previously extended] [further] extension of time.

Any delay [beyond any peiod previously granted] is considered solely due to circumstances or default for which you as contractor are solely responsible under the terms of the contract.

This notice is given pursuant to Clause 28(2) of the conditions.

Yours faithfully

Superintending Officer

Reference: **GC/SO/38**

To: The Contractor

Dear Sir

Notification of Intention to Deduct Liquidated Damages

We note that the Works/that section of the works being ....
.... was/were not completed by the due date for completion [as extended] and continued so uncompleted until ....
....

As a consequence we are notifying the Authority of his right to levy liquidated damages for delay amounting to £................................ pursuant to Clause 29(1) of the conditions. We anticipate that the Authority will deduct and retain such sum from any further sum otherwise payable to you.

Yours faithfully

Superintending Officer

CHAPTER 12

Subletting and Defects Liability

Clause 30(1)—Subletting

The Contractor shall not sub-let any part of the Contract without the previous consent in writing of the SO.

Comment

1. As under both the JCT and ICE forms of contract, permission must be obtained prior to sub-letting any part of the works.

2. Where part of the works has been sub-let with the sanction of the SO, the terms of the sub-contract must incorporate the following provisions:

(i)   a right to determine the sub-contract pursuant Clause 44(6);

(ii)   that where materials are brought onto the site by a sub-contractor they shall vest in the main contractor;

(iii)   that provision shall be included that will enable the main contractor to fulfil his contract obligations, particularly under the following clauses:

(a)	3(2)	things brought onto the site not to be removed without the consent of the SO,
(b)	4(4)	on the completion or the early determination of the contract all documents shall be returned to the SO,
(c)	13(2)	the SO shall have right of inspection on the premises of the sub-contractor,
(d)	13(3)	the SO shall be entitled to have tests made on sub-contract materials,
(e)	35	where the works are being carried out within the boundaries of a government establishment the sub-contractor shall comply with the rules and regulations of the establishment,
(f)	56	the sub-contractor's personnel shall be subject to sanctions relating to their admission to the Site,

(g)	57	the requirements related to the provision of passes to the site of the works,
(h)	58	photographs not to be taken unless sanctioned by the SO,
(j)	27	the sub-contract not to be further assigned without the permission of both the contractor and the SO,
(k)	36	the right of the SO to require the removal and replacement of any person below the grade of foreman employed by the sub-contractor,
(l)	55	penalty relating to corrupt gifts and the payment of commission,
(m)	59	abiding by the requirements of the Official Secrets Act,
(n)	11G(1)(2)	the incorporation of the fluctuation clause related to labour-tax matters.

3. Sub-contracted works shall not be further sub-let without the consent of the contractor, and presumably the sanction of the SO, although no specific mention is made of this latter requirement.

Clause 31(1)—Objection to the Appointment of a Nominated Sub-Contractor or Supplier

No person against whom the Contractor shall make reasonable objection shall be employed as a nominated sub-contractor or nominated supplier in connection with the Works.

Comment

1. The right to objection to the appointment of a nominated sub-contractor or supplier is a right which exists not only under these conditions but also under the JCT and ICE forms.

However, whilst the right of objection itself is similar, these conditions unlike the other two are studiously silent as to what shall happen should the contractor raise reasonable objection.

2. The burden that would be suffered by the Contractor in the event of a failure of a nominated sub-contractor is detailed in sub clause (3), where it is stated that the Authority would not make a substitute nomination:

'The Contractor shall make good any loss suffered or expense incurred by the Authority by reason of any default or failure, whether total or partial, on the part of any sub-contractor or supplier.'

Should there be failure on the part of the sub-contractor, the contractor would have to make his own arrangements to ensure the sub-contracted works were completed to the terms of the specification by whatever resources he might consider appropriate. This would be totally at his own cost so far as additional expense is concerned over and above that that would have been incurred in relation to the original sub-contractor. It is therefore crucially important that objection under this form be made at the appropriate time, should ground exist to doubt the financial stability or capability of the firm being nominated.

3. In view of the extremely onerous position in which the Contractor will find himself should a nominated sub-contractor be terminated, as is indicated in Clause 38(5), proper attention to the right of objection is probably one of the most significant commercial considerations under these conditions. Very commonly the adequate and economical execution of the works will depend on the competent programmed performance of the various specialist sub-contractors. If the contractor finds himself encumbered with a group of managerially inept and financially insecure sub-contractors, he will have little chance of performing his contract adequately and no chance at all of performing it profitably should any one of the significant sub-contractors fail.

4. Regrettably, postulation of the grounds for reasonable objection will almost inevitably give rise to a degree of acrimony, since ground for objection is likely to be lodged under one or more of the following headings:

(a) that the sub-contractor refuses to enter into a sub-contract that would satisfactorily indemnify the main contractor against the same liabilities in respect of the sub-contract works as he has to undertake in respect of the main contract works;

(b) that the sub-contractor is believed to be financially insecure;

(c) that previous experience of working with the sub-contractor on comparable works has shown him to be incapable of handling work of the quantity and complexity called for in the scheme;

(d) that his availability of skilled labour will be inadequate for the scheme;

(e) that he has insufficient managerial capability for the works concerned;

(f) that he is unable to comply with the programme requirements of the main contract.

Remarks of this nature, which may of necessity have to be made in correspondence with the SO, must obviously be well researched and properly founded with background evidence which the contractor must be prepared to produce. However, bearing in mind the consequences to the contractor's own business if he is landed with an inadequately performing sub-contractor, or even worse, one that fails in liquidation, it will not pay him to be too squeamish.

Remember—*object before the sub-contract is entered into with the nominated firm, since after nomination all recourse to the Authority for recompense and financial assistance in the event of failure of the firm concerned will have been lost.*

5. It should also be appreciated that whilst no reference is made in the conditions to the sub-contractor's right to object to being nominated to a main contractor, embarrassing though this may be, such right equally exists. At the stage when a contractor is required to place a nomination with a sub-contractor no contract exists between the two parties and therefore the sub-contractor can merely refuse to accept the offer being made to him by the contractor.

The contractual situation with nominated sub-contractors as regards offer and acceptance is that the contractor offers the sub-contract work to the sub-contractor at a certain price and under certain terms and conditions which the sub-contractor may or may not accept. If he accepts, a sub-contract comes into being, and if he does not, no sub-contract exists.

Clause 31(2)—Responsibility for Sub-Contractors or Suppliers

The Contractor shall be responsible for any sub-contractor or supplier employed by him in connection with the Works whether he shall be nominated or approved by the Authority or the SO, or shall be appointed by the Contractor in accordance with the directions of the Authority or the SO or otherwise.

Comment

1. The clause is here making the standard stipulation included in all the standard forms of contract, 'that no privity of contract shall exist between the Employer and a sub-contractor or supplier whether nominated or otherwise under the main contract'.

2. If therefore one or other of the sub-contract obligations between the main contractor and the sub-contractor is unfulfilled, the contracting parties under the sub-contract must have recourse one to the other for their remedy. Depending on the nature of the obligation that has not been fulfilled, it may or may not be that recourse may eventually be made by the main contractor to the employer for recompense if such failure is itself the result of a breach by the employer of one or other of the main contract obligations.

Clause 32—Defects Liability

(1) This Condition applies:

(a)   to any defects (excluding any defects specified in sub-paragraph (b) below) which may appear within the maintenance period specified in the Abstract of Particulars in respect of the Works; and

(b)   to any defects in any sub-contract works in respect of which a separate sub-contract maintenance period is specified in the Abstract of Particulars, which may appear within the appropriate sub-contract maintenance period,

being (in either case) defects which arise from any failure or neglect on the part of the Contractor or any sub-contractor or supplier in the proper performance of the Contract or from frost occuring before completion of the Works.

(2) The Contractor shall make good at his own cost to the satisfaction of the SO any defect to which this condition applies:

Provided that he shall not be required to make good at his own cost any damage by frost which may appear after completion unless the cause of such damage arose at a time before completion.

(3) In case of default the Authority may provide labour and/or any things necessary, or may enter into a contract or contracts, in order to repair and make good any defects to which this Condition applies, and all costs and expenses consequent thereon shall be borne by the Contractor and shall be recoverable from the Contractor by the Authority.

(4) In the case of any defects specified in paragraph (1)(b) of this Condition which have been made good, the provisions of paragraphs (2) and (3) of this Condition shall apply to the sub-contract works which have been made good until the expiration of either the sub-contract maintenance period or a period of six months from the date of making good (whichever is the later).

Comment

1. Under these conditions the defects liability period commences from the date of the certificate issued under Clause 42(1) that certifies the date on which the works were completed. The definition of completion under this form is rather more onerous than under either the JCT or ICE Conditions since it will be noted that under Clause 6 it talks of 'the whole of the Works shall be completed by the date for completion' and under Clause 28(1) requires 'the Works cleared of rubbish and delivered up to his satisfaction on or before the date for completion'.

2. The degree of total completion implicit in these clauses, if enforced, will mean that a stated contract period will be effectively shorter than under either the JCT or ICE Conditions in that the state of completion of the works will need to be more advanced at completion than in the case of either of the other two forms. It is inevitable that if the SO has strict regard to the form of words used in the contract, the obtaining of a certificate of completion may prove to be a somewhat difficult process.

3. The conditions do not make formal provision for the issue of a schedule of defects, although presumably such could and should be issued under the terms of Clause 7(1)(j), under which instructions may be issued regarding the amending and making good of defects under Clause 32. Unfortunately this clause indicates no time-scale for the provision of a schedule of defects and the contractor must rely on arguing that the schedule must be received at a time which, in relation to the duration of the defects liability period, is not unreasonably close to or distant from the end of that period. The time of issue should allow the contractor sufficient time to make good the defects noted within the total time-scale of the defects liability period stipulated and at its end to have issued to him a certificate under Clause 42 certifying when the works were in a satisfactory state.

4. As under all defects liability clauses, the Contractor must take care to ensure that he does not inadvertently carry out work which is not a defect but merely a variation purporting to be a defect.

The schedule of defects should be examined to identify those items that are genuine defects as distinct from those that are purported defects. To achieve this the schedule should be split in the following way:

(a) genuine defects;
(b) defects arising from fair wear and tear by the user;
(c) defects arising from some inadequacy in the specification or design with which, in accordance with the terms of the contract, the contractor has complied;
(d) defects which on examination are shown to be a variation.

Since the contractor is required to make good at his own cost to the satisfaction of the SO any defect to which the clause applies, namely category (a) only, care should be taken to ensure that an instruction to carry out works in the other three categories is received under Clause 7(1)(a) and not under (j).

5. It should be noted that the clause envisages the possibility of the defects liability period in respect of certain of the sub-contracted works being of a different duration to that of the main contract. In consequence the contractor may need to have supervisory resource associated with the contract for a longer period than would have been apparent by superficial inspection of the abstract of particulars.

STANDARD LETTERS

Reference: **GC/32**

To: The SO

Dear Sir

Request for Sanction to Sub-Let

We request consent to sub-let the following item/items of work to the undermentioned firm/firms pursuant to Clause 30(1) of the Conditions and undertake to ensure that the terms of our sub-contracts incorporate the provisions of Clauses 30(2) and (3) as appropriate.

Firm   *Item of Work*

....

....

We believe the firm/firms to have the degree of competence and resource appropriate to the satisfactory and timely execution of the work/works involved.

Yours faithfully

Contractor Limited

Reference: **GC/33**

To: The SO

Dear Sir

Objection to the Nomination of a Nominated Sub-Contractor/Nominated Supplier

We refer to your instruction dated .... nominating
.... for
.... and regret to inform you of our
objection to the nomination pursuant to Clause 31(1) of the conditions
on the ground that:

( ) the nominated firm declines to enter into a sub-contract/contract of
 sale that will satisfactorily indemnify ourselves as main contractor
 against the same liabilities in respect of the sub-contract
 works/supplies as we have to undertake to the Authority in respect of
 the main contract works;
( ) the nominated firm is believed to be financially insecure;
( ) the nominated firm has in our previous experience shown itself
 incapable of handling work of the quantity and complexity called for
 in the present scheme;
( ) the nominated firm has an insufficient availability of skilled labour
 for the project;
( ) the nominated firm lacks the managerial capability to handle the
 works concerned;
( ) the nominated firm is unable to comply with the programme
 requirements of the main contract [previously notified to the
 Authority].

Your further instructions are therefore sought on this matter.

Yours faithfully

Contractor Limited

Reference: **GC/34**

To: The SO

Dear Sir

Clarification of Responsibility for the Costs of Defects

We refer to the schedule of defects notified by the SO pursuant to Clause
32 of the conditions and question whether the following items

(i) ....
(ii) ....
(iii) ....
(iv) ....

are attributable to any failure or neglect on our part or on the part of any of our sub-contractors or suppliers in the proper performance of the Contract [or from frost occurring before completion of the works].

Believing these items of purported defect to have arisen from other causes, we request their value be ascertained and paid for as if it were variation work pursuant to Clause 9(1) of the conditions.

Yours faithfully

Contractor Limited

Reference: **GC/35**

To: **The SO**

Dear Sir

Request for a Schedule of Defects

We note that on .... the maintenance period specified in the Abstract of Particulars ends.

To enable us to complete our contract liabilities efficiently to the benefit of both the Authority and ourselves we would ask for the early issue of the schedule of defects.

Yours faithfully

Contractor Limited

Reference: **GC/SO/39**

To: The Contractor

Dear Sir

Approval of Request for Sanction to Sub-let

We refer to your letter dated .... reference
.... requesting permission to sub-let the undernoted work.

Item of Work	*Firm*
....	....
....	....
....	....

Our consent is given pursuant to Clause 30(1) of the conditions.

Yours faithfully

Superintending Officer

Reference: **GC/SO/40**

To: The Contractor

Dear Sir

Denial of Request for Sanction to Sub-let

We refer to your letter dated .... reference
.... requesting permission to sub-let the undernoted work.

Item of work	*Firm*
....	....
....	....
....	....

We regret that we are unable to give our consent to your request pursuant to Clause 30(1) of the Conditions.

[You will appreciate that this denial of consent is a matter of some delicacy and whilst not wishing to put our reasons for the denial in writing we are prepared to discuss the matter with your representative.]

[Our reason for this decision is that...

....]

Yours faithfully

Superintending Officer

Reference: **GC/SO/41**

To: The Contractor

Dear Sir

Notification of Appointment of Nominated Sub-Contractor/Nominated Supplier

We hereby direct you to place an order with ....
.... for .... in accordance with the enclosed documentation. Such firm will be nominated sub-contractor/supplier under the terms of the contract.

This direction is given pursuant to Clause 31(2) of the conditions.

Yours faithfully

Superintending Officer

Reference: **GC/SO/42**

To: The Contractor

Dear Sir

Issue of Schedule of Defects

We issue herewith a schedule of defects of various items which are required to be made good pursuant to Clause 32 of the conditions.

Should you consider any of the notified items to be attributable to fair wear and tear or to have arisen despite, in your opinion, compliance with the contract in respect of materials and workmanship, please notify us accordingly within 14 days.

If in our opinion such defects are so caused, the value of work entailed in making good will be ascertained and paid for as additional work under the terms of Clause 9.

All items of actual defect, whether by way of additional work or otherwise, are required to be made good prior to the issue of the 'Certificate of Satisfactory State' under Clause 42(1) of the conditions, the issue of which is a condition precedent to final payment.

Yours faithfully

Superintending Officer

CHAPTER 13

Contractor's Employees and Measurement of the Works

Clause 33—Contractor's Agent

The Contractor shall employ a competent agent to whom directions may be given by the SO. The agent shall superintend the execution of the Works generally with such assistance in each trade as the SO may consider necessary. Such agent shall be in attendance at the Site during all working hours except that when required to do so he shall attend at the office of the SO.

Comment

1. The Clause contains an interesting power which in practice is fortunately rarely exercised. In theory the SO may require that the agent shall have certain specialist trade supervision to the extent that the SO may consider it necessary.

2. An over-zealous SO could result in an extremely contentious situation arising on a contract. Obviously when a contractor tenders, he tenders on the assumption that the works will be the subject of a certain level of supervisory resource, which bearing in mind the duration and characteristics of the works, he will consider adequate. Should the SO subsequently issue instructions requesting an additional level of resource beyond that reasonably contemplated by an experienced Contractor and thereby cause direct loss and expense, doubtless the contractor would seek to recover the additional cost by a claim.

Clause 34—Daily Return

The Contractor's Agent shall provide the SO each day with a distribution return of the number and description of workpeople employed on the Works.

Comment

1. These daily returns are designed primarily to enable the Authority to monitor the contractor's labour force in relation to the achievement of the work programme for the project.

2. In practice the returns provide an invaluable source of substantiating data both to the contractor in preparing a claim and to the Authority in assessing its justification. The return, if properly prepared, should provide a day-to-day record of:

(a) where the labour force was working;
(b) the numbers of that labour force;
(c) the trades of the people concerned.

Condition 35—Contractor to Conform with Regulations

Where the Works are to be executed within the boundaries of a Government Establishment, the Contractor shall comply with those Rules and Regulations of the Establishment which are described in the Contract.

Comment

1. It is obviously perfectly reasonable that the contractor be required to comply with specific rules and regulations providing they have been detailed in the tender documents and the contractor given the opportunity to reflect their financial impact in his price.

2. Should the contractor be required to comply with restrictive regulations that had not been highlighted in the tender documents and such compliance causes additional loss and expense that could not reasonably have been contemplated by an experienced contractor at time of tender, then such additional cost would be a matter for claim.

In this regard it is most important that the contractor checks that the documents on the basis of which he is required to enter into contract are in fact precisely those on which he tendered. As under all other conditions of contract, it is all too easy to find that liabilities have been slipped into the contract between the time of tender and the time of acceptance. If that is done, the Authority is making a counter-offer and it is entirely up to the contractor as to whether he accepts the counter-offer. Having done so, however, he would be landed with the consequences (subject only to possibly contested legal proceedings) if he had not spotted any problem and picked it up accordingly.

3. It is likely that any instruction given to comply with a regulation that had not been detailed in the tender documents and was subsequently detailed in the contract documents would be given under Clause 7(1)(m), under which the SO is empowered to issue instructions on any other matter as to which it is necessary or expedient. So far as compliance will have caused additional cost, the claim would then be lodged under Clause 9(2).

Clause 36—Replacement of Contractor's Employees

(1) The SO shall have power to require the Contractor, subject to compliance with any statutory requirements, immediately to cease to employ in connection with the Contract and to replace any foreman or person below that grade whose continued employment thereon is in the opinion of the SO undesirable.

(2) The Authority shall have power to require the Contractor immediately to cease to employ in connection with the Contract and to replace any person above the grade of foreman, including the Contractor's Agent, whose continued employment in connection therewith is in the opinion of the Authority undesirable.

(3) Any decision of the Authority or the SO under this Condition shall be final and conclusive.

Comment

1. It is to be noted that the powers under the contract requiring removal of a contractor's employee are couched in two stages. The SO has such power at foreman level and below, while only the Authority can exercise that power over a higher grade of staff.

2. The clause makes it quite clear that the contractor has no right of appeal against such a requirement, as we are told 'any decision of the Authority or the SO . . . shall be final and conclusive'.

3. While this must mean that no argument can be brooked in regard to compliance with the direction, it would not necessarily appear to debar the Contractor from seeking compensation were it to involve him in loss and expense beyond such as reasonably could have been foreseen by an experienced contractor at time of tender. Were this to be done, the direction would need to be treated as an instruction given under Clause 7(1)(m), triggering a right to compensation under Clause 9(2).

4. If compensation is to be sought for the removal and replacement of an employee, the reason for the removal must be due to a cause that is not attributable to failure on the part of the individual to carry out his duties appropriately, or that relates to some regulation of which the contractor had been given notice at time of tender.

Clause 37(1)—Attending for Measurement

The Contractor's representative shall from time to time, when required on reasonable notice by the Quantity Surveyor, attend at the Works to take jointly with the Quantity Surveyor any measurements of the work executed that may be necessary for the preparation of the Final Account. Any such measurements when ascertained and any differences arising therefrom shall be recorded in the manner required by the Quantity Surveyor. The Contractor shall without extra charge provide assistance with every appliance and other thing necessary for measuring the work. If the Contractor's representative fails to attend when so required, the Quantity Surveyor shall have power to proceed by himself to take such measurements.

Comment

1. Unlike the ICE Conditions of Contract, this form specifically recognises the contractual existence of the quantity surveyor. It is the quantity surveyor who will take the necessary measurements of the work and prepare the final account, acting on behalf of the SO.

2. The conditions clearly contemplate that the measurements required shall be taken progressively and agreed between the parties as work proceeds. This is an admirable intent and one which if regularly followed would avoid so much of the financial problems that occur not only under these conditions, but all conditions.

3. The clause recognises the possibility of differences of opinion arising over measurement although it does not establish the way in which the differences are to be resolved. However, the contractor should recall his right under Clause 5(2) to have corrected any error in description or in quantity in the bills of quantities or any omission therefrom. The guiding light as to what is, or is not, a mistake will of course be the stipulated Method of Measurement in accordance with which the bills are supposed to have been prepared as indicated in Clause 5(1).

4. The onus rests on the contractor to ensure that his representative is available at reasonable notice to work with the quantity surveyor in taking the measurements referred to. In the event of the contractor's representative failing to attend, the quantity surveyor appointed under the contract is empowered to proceed with the measurements alone and the measurements as taken shall be deemed to represent the works executed.

5. The clause talks of 'any measurements' in relation to the preparation of the final account. The term applies equally therefore to work that is actually site-measured and to work measured from the drawings. In practice, of course, the preponderance of work will be measured from drawings rather than be based on site measurement as such.

6. Although the contractor is required without extra charge to provide such assistance and appliances as are necessary for measuring the work, if the contract had been represented as a comprehensively designed and firm contract on the basis of bills that were not designated as provisional or approximate, and it was subsequently found that the works were the subject of an inordinate level of variation, the contractor might well make the provision of measuring resource, to the extent that it was unreasonable and outside the contemplation of the contract, a subject of claim.

It must be appreciated that firm bills of quantities are of themselves a representation as to the design state of the works, and if in reality the supposed-firm bill is really a bill of approximate quantities masquerading as something that it is not, the contractor has ground for complaint and might treat the matter as a misrepresentation under the Misrepresentation Act 1967.

Clause 37(2)—Provision of Information

The Contractor shall provide to the Quantity Surveyor all documents and information necessary for the calculation of the Final Sum (including for Conditions 9, 20 and 53) certified in such manner as the Quantity Surveyor may require.

Comment

1. The words in brackets 'including for Conditions 9, 20 and 53' make it quite clear that the contractor must produce such substantiating information in regard to claims as the quantity surveyor calls for.

2. The more routine type of documentation called for by the clause would be of the following nature:

(a) Increased cost claims substantiated by wage sheets and invoices or calculation by way of the NEDO formula as appropriate,
(b) daywork sheets,
(c) authorised accounts of nominated sub-contractors and suppliers,
(d) additional information required to enable star rates to be agreed.

STANDARD LETTERS

Reference: **GC/36**

To: The SO

Dear Sir

Labour Return

We attach our daily labour return for .... pursuant to Clause 34 of the conditions in the form requested showing:
(a) where the labour was working;
(b) the numbers involved; and
(c) the trades of the people concerned.

Yours faithfully

Contractor Limited

Reference: **GC/37**

To: The Quantity Surveyor

Dear Sir

Submission of Documents in Connection with the Final Account

We enclose herewith the documents listed below in connection with the preparation of the Final Account as required by Clause 37(2) of the Conditions.

  (i)   ....
  (ii)   ....
 (iii)   ....
 (iv)   ....

Yours faithfully

Contractor Limited

Reference: **GC/SO/43**

To: The Contractor

Dear Sir

Request for Returns of Labour, Plant and Related Matters

You are required to provide a daily labour return covering both your own resources and those of any approved sub-contractor pursuant to Clause 34 of the Conditions.

Such returns shall be made in a form approved by the SO.

Yours faithfully

Superintending Officer

Reference: **GC/SO/44**

To: The Contractor

Dear Sir

Superintending Officer's Request for Replacement of Contractor's Employees

We regret that we must direct you to remove Mr .... from the works pursuant to Clause 36(1) of the Conditions on the ground that he has misconducted himself/is incompetent/negligent in performance of his duties/has failed to conform with the particular provisions with regard to safety.

[Appreciating this to be a matter of some delicacy, we have refrained from stating our detailed reasons for our directive in this notice but will be prepared to discuss the matter with a senior and accredited representative from your Company.]

Yours faithfully

Superintending Officer

Reference: **GC/SO/45**

To: The Contractor

Dear Sir

Authority's Request for Replacement of Contractor's Employees

We regret that we must withdraw our approval of Mr .... as your competent and authorised agent/................. pursuant to Clause 36(2) of the conditions.

We direct that he be withdrawn from the site immediately/as soon as can reasonably be arranged without detriment to the works.

[Appreciating this to be a matter of some delicacy, we have refrained from stating a reason for our directive in this notice but will be prepared to discuss the matter with a senior and accredited representative of your Company.]

Yours faithfully

Authority

Reference: **GC/SO/46**

To: The Contractor

Dear Sir

Notification to Attend for Measurement

We refer to the requirements of Clause 37(1) and list below a schedule of dates to be left clear for measurement purposes:

  (i)   ....

  (ii)   ....

  (iii)   ....

  (iv)   ....

Yours faithfully

Superintending Officer

CHAPTER 14

Nominated Sub-Contractors and Suppliers, Provisional Sums and Provisional Quantities

Clause 38(1)—Definition of Prime Cost

The words 'Prime Cost' or the initials 'PC' applied in the Contract to any work to be executed or any things to be supplied by a sub-contractor or supplier shall mean that in respect of such an item the sum to be paid by the Authority shall be the sum (inclusive of proper charges for packing, carriage and delivery to the Site) due to the sub-contractor or supplier, after adjusting in respect of over-payment or over-measurement or otherwise and after deduction of all discounts obtainable for cash in so far as such discounts exceed $2\frac{1}{2}\%$ and of all trade discounts, rebates and allowances.

Comment

1. The sum of money covered by a prime cost in the bills of quantities will relate to a specific item of work or supply of goods by a nominated sub-contractor or nominated supplier. Both nominated sub-contractors and nominated suppliers under this form are inclusive of a $2\frac{1}{2}\%$ main contractor's cash discount for prompt payment, whereas under the JCT form the discount provision is $2\frac{1}{2}\%$ in respect of a nominated sub-contractor and 5% in respect of a nominated supplier.

2. The clause makes provision for the amount to be paid to a nominated firm being adjusted in respect of over-payment or over-measurement. Contractors should therefore be very wary of making interim payments to nominated firms purely on the say-so of amounts certified by the SO without further checking. Recovery from the firms concerned, in the event of over-payment, may prove extremely difficult, whilst the amount paid to the main contractor may be corrected regardless of whether or not he is able to make recovery from the nominated firm.

Clause 38(2)—Contractor's On-Cost on Prime Cost Items

The Contractor shall also be entitled to payment for fixing in accordance with the rates included in the Bills of Quantities or the Schedule of Rates and to Contractor's profit where applicable. The payment for fixing shall cover unloading, getting in, unpacking, return of empties and other incidental expense. The Contractor's profit at the rate included in the Bills of Quantities or the Schedule of Rates shall be adjusted pro-rata on the prime cost excluding any alterations in that prime cost due to the operation of any conditions incorporated in the sub-contract pursuant to Condition 30(3).

Comment

1. In the case of a *nominated supply item* the items to be priced by the Contractor in connection therewith will be as follows:

(a) An item for fixing with sufficient descriptive content to be able to assess adequately the nature of the fixing requirement. It should be noted that whilst 'unloading, getting in, unpacking, return of empties and other incidental expense' would be included in the contractor's fixing item, the proper charges for packing, carriage and delivery to the site would be included in the price for the nominated supply item. Were such packing, carriage and delivery not to have been included in the nominated supply price, then it should be added to the sum concerned as a matter of contractual right so far as the contractor is concerned.

(b) Over and above the fixing item the contractor will have an opportunity to price for profit on the amount of the sum included in the Contract. The profit allowance will subsequently be adjusted pro-rata to any change in the sum concerned (other than a change that arises purely as a result of an adjustment arising from a change in a labour tax matter).

2. In the case of a *nominated sub-contractor item*, since the bills of quantities for works carried out under this form of contract will usually have been prepared in accordance with the Standard Method of Measurement of Building Works, the type of ancillary items to be priced will be of the following nature:

(a) General attendance on nominated sub-contractors shall be given as an item in each case and shall be deemed to include only allowing use of standing scaffolding, mess rooms, sanitary accommodation and welfare facilities; providing space for office accommodation and for storage of plant and materials; providing light and water for their work; clearing away rubbish.

(b) Special attendance on nominated sub-contractors shall be given as an item in each case giving particulars (e.g. unloading; storing; hoisting; placing in position; providing power; providing special scaffolding).

(c) Builders' work in connection with work by nominated sub-contractors shall be given in accordance with the appropriate rules stated throughout the Standard Method of Measurement appropriate to the work concerned. It is desirable that such work be grouped together under a heading in each appropriate section of the bill.

(d) In addition it will be noted pursuant to Clause 38(2) that the contractor is to have been given an opportunity to price the profit element on the amount of the sub-contract as a separate item, again conditional on that sum containing no allowance for increased costs pursuant to changes resulting from a labour-tax matter.

Clause 38(3)—Increase or Decrease in a Prime Cost Sum

Any increases or decreases in the prime cost sums included in the Contract resulting from these adjustments shall be added to or deducted from the Contract Sum. The Contractor shall produce to the Quantity Surveyor such quotations, invoices and bills (properly receipted) as may be necessary to show the actual details of the sums paid by the Contractor.

Comment

1. Although the amount of a nominated sub-contract item covered by a 'prime cost' will normally have been agreed by the appointed quantity surveyor, the contractor will still be required to produce authenticating documentation to demonstrate that the sum agreed has in fact been the sum paid by the contractor.

2. The purpose of the sub-clause would appear to be to prevent any collusion between the sub-contractor and the main Contractor whereby a greater discount than is allowed under the terms of the Contract might be available to the Contractor.

Clause 38(4)—Nomination of a Sub-Contractor or Supplier

All prime cost items shall be reserved for the execution of work or the supply of things by persons to be nominated or appointed in such ways as may be directed by the Authority or the SO and the Contractor shall not order work or things under such items without the written instruction of the SO or consent of the Authority. The Authority reserves the right to order and pay for all or any part of such items direct and to deduct the sums included therefor from the Contract Sum less an amount in respect of Contractor's profit at the rate included in the Bills of Quantities or Schedule of Rates adjusted pro-rata on the amount paid direct by the Authority.

Comment

1. Under this clause the SO may only instruct the contractor in writing in relation to the carrying out of works that are the subject of nomination. It must be appreciated that these instructions relate not only to the initial nomination but to all subsequent instructions that may be issued in connection with the sub-contract of the scope and nature defined in Clause 7. It is essential therefore that the contractor passes on all instructions to the various sub-contractors in writing and equally that they for their part should never take or act on an instruction other than one in writing given by the contractor.

2. These conditions, unlike earlier versions of the Government form of contract, now enable the Authority to pay a nominated sub-contractor direct and to deduct such monies from the amounts otherwise due to the contractor. In this respect the conditions accord with the arrangements of both the JCT and ICE forms.

If such direct payment has occasion to be made, the Contractor's entitlement to profit on the amount of the sub-contract is adjusted pro-rata to the amount of payment that has been made direct. This adjustment in respect of the contractor's financial additions to the sub-contract applies only to profit and not to other ancillary items such as attendance.

Clause 38(5)—Termination of a Nominated Sub-Contract

In the event of termination of a sub-contract to which this Condition applies, the Contractor shall, subject to the consent in writing of the Authority, either select another sub-contractor or supplier to undertake or complete the execution of the work or the supply of the things in question, or himself undertake or complete the execution of that work or the supply of those things and the Authority shall pay the Contractor the sum which would have been payable to him under paragraph (1) of this Condition if termination of the said sub-contract had not occurred, together with any allowances for profit and attendance which are contained in the Bills of Quantities.

Comment

1. The point has already been made in discussing the right of the contractor to object to the nomination of a nominated sub-contractor that in the event of the termination of the sub-contract the contractor will find himself liable for the additional costs of ensuring that the sub-contracted works are completed by some other sub-contractor or by the directly employed labour of the contractor. The most likely single cause of termination of the sub-contract will be the bankruptcy of the sub-contractor concerned and it is therefore crucially important that the contractor make adequate financial enquiry as to the stability and financial backing of the firm being nominated before being prepared to accept the nomination.

2. In the past, the normal procedure in the event of a termination of the contract of a nominated sub-contractor was for the employer to re-nominate. This entailed the setting up of a totally new sub-contract governed by such prices as it was possible to obtain. In practice this often meant that the employer would pay substantially more for the work and furthermore usually suffered delay to his contract.

3. It is interesting to note that whilst the ICE Conditions have evolved to a much more equitable approach to the problems of a nominated sub-contractor, so far as the main contractor is concerned, in the case of the Government form of contract, the movement appears to have been in exactly the opposite direction.

It is all too easy to imagine the financial dog-fight that would occur between a main contractor and any major replacement sub-contractor who takes up the work of a nominated sub-contractor following the latter's bankruptcy and subsequent termination. The sub-contractor in these circumstances will inevitably be faced with a more expensive job than the content of the original sub-contract.

4. The work he will be undertaking will usually be in a partially completed stage and possibly subject to deterioration and neglect arising from the passage of time between the bankruptcy of the original sub-contractor and the stage at which he, as a successor, is able to commence work. The main contractor, on the other hand, will be resisting every aspect of increased cost in order to limit the level of loss he will inevitably be making, further compounded by the fact that appreciable main contract overrun, for which recompense would not appear to be available, will also be occurring.

To get this matter in perspective, let us imagine the case of a £1,000,000 building scheme where, at an early stage of carrying out his work, the nominated sub-contractor for mechanical services went into liquidation. It is quite possible, in a reasonably sophisticated building, for the mechnical services to represent 30% of the total value of the work. In the circumstances we have been outlining, getting the mechanical services completed might cost half as much again as if it had been completed by the original sub-contractor. In terms of our example this equates to 15% of the cost of the job, which is likely to be more than the whole of the main contractor's overhead and profit cover. Therefore, by having failed to make objection to nomination at the appropriate time, the contractor is faced with the crippling situation of seeing the whole of his overhead and profit cover wiped out at a stroke.

5. It would appear to the writer that this provision, under the Government form of contract, might well trigger a chain of bankruptcy in which the financial burden placed on the main contractor might bring him down and carry with him the second sub-contractor.

While prudently run major contracting organisations should be able to avoid most of these potential hazards, and in the case where they might be eventually caught, survive the ordeal, there is no doubt that this particular piece of contract draftsmanship is an arrangement that can only cause considerable hardship to the industry and eventually its clients.

Clause 39—Provisional Sums and Provisional Quantities

The full amount of the provisional sums included in the Contract and the net value annexed to each of the provisional items inserted in the Bills of Quantities shall be deducted from the Contract Sum and the value of work ordered and executed thereunder shall be ascertained as provided by Condition 9(1) or 10 as the case may be. No work under these items shall be commenced without instructions in writing from the SO.

Comment

1. The SO's instructions in relation to the expenditure of provisional sums or the use of provisional quantities would presumably be issued under Clause 7(1)(m), under which an instruction may be issued on any other matter for which it is necessary or expedient. When the execution of work, the subject of a provisional sum, has been valued under the terms of Clause 9(1) and causes the contractor a level of expense beyond that reasonably to have been provided for in or contemplated by the contract, any such additional amount may be the subject of claim pursuant to Clause 9(2).

2. In general, the provisional sums referred to will be sums of money incorporated in the contract documents to cover work, the scope of which at time of tender may not yet have been defined and the manner of its contractual execution not yet decided.

3. In the case however of provisional quantities it is likely that the general nature of the works required will have been established but that its detail will not have been sufficiently defined to enable firm quantities to be measured. Typically, provisional quantities might relate to such items as foundation works, builders' work in connection with services sub-contracts and possibly external works, where rather as under a civil engineering contract it might not be possible to establish the full scope and nature of the works until the ground was opened up or a particular specialist appointed.

4. In order to avoid delay associated with works related to provisional sums and provisional quantities, it obviously behoves the contractor to request an instruction relating to the items concerned well before the stage when the contractor would need to carry out the work. It should be appreciated that the very reason for the items appearing on a provisional basis in the first place may well have been because the definitive information relating to them did not exist. Obviously in those situations it is reasonable to conclude that the SO may require time in order subsequently to produce the information. To sit by and merely wait for a totally unsolicited instruction, while possibly not being contrary to the wording of the contract, would be commercially extremely foolhardy.

STANDARD LETTER

Reference: **GC/38**

To: The SO

Dear Sir

Request for an Instruction Relating to Provisional Sum/Provisional Quantities

We request your written instruction pursuant to Clause 39 of the conditions relating to the undermentioned provisional sums and provisional items contained in the bill of quantities:

  (i) ....
  (ii) ....
 (iii) ....
 (iv) ....

We understand that no work involving these items may be commenced without the appropriate instruction [and particularly in regard to .... such instruction is a matter of urgency].

Yours faithfully

Contractor Limited

CHAPTER 15

Payments and Certificates

Clause 40—Payments on Account

(1) The Contractor shall be entitled to be paid during the progress of the execution of the Works 97 per cent of the value of the work executed on the Site to the satisfaction of the SO and the Authority shall accumulate the balance as a reserve.

(2) The Contractor shall also be entitled to be paid during the progress of the execution of the Works 97 per cent of the value of any things for incorporation which are in the opinion of the SO in accordance with Contract and which have been reasonably brought on the Site and are adequately stored and protected against damage by weather or other causes, but which have not at the time of the advance been incorporated in the Works. When any things on account of which an advance has been made under this paragraph are incorporated in the Works the amount of such advance shall be deducted from the next payment under paragraph (1) of this Condition.

(3) The Contractor may at intervals of not less than one month submit claims for payment of advances on account of work done and of things for incorporation which have been delivered. Such claims shall be supported by a valuation of the work done and of things so delivered, which valuation shall be made on the basis of the rates in the Bills of Quantities or on the Schedule of Rates or, where such rates are not applicable, on the appropriate alternative basis of valuation set forth in Conditions 9(1) or 10. When the valuation has been agreed by the SO, the SO shall certify the sum to be paid by way of advance:

Provided that if the Contract Sum exceeds £100,000 there shall be paid to the Contractor on his application at the end of the second week in each monthly period an interim advance on account of the further work done or things for incorporation supplied since the date of the last valuation. The amount of any such interim advance shall be an approximate estimate only and the decision of the SO in regard thereto shall be final and conclusive.

(4) Any sum agreed to be credited by the Contractor for old materials shall be deducted from the first or any subsequent advance.

(5) Without prejudice to the Contractor's entitlement to an increase in the Contract Sum under Conditions 9(2) or 53(1), to the Authority's entitlement to a decrease in that sum under Condition 9(2) or to the amount of any such increase or decrease, where the SO is of the opinion that there is to be any such increase or decrease he shall decide an amount to be added on account of that increase, or to be deducted on account of that decrease, to or from any sum falling to be paid to the Contractor under paragraph (3) of this Condition and a sum equal to the amount so decided shall be added or deducted, as the case may be, to or from a sum falling to be so paid.

(6) Before the payment of any advance or the issue of any final certificate for payment the Contractor shall, if requested by the SO, satisfy him that any

146

amount due to a sub-contractor or supplier of things for incorporation which is covered by any previous advance has been paid. In any case where the SO is not satisfied as aforesaid:

(a) the Authority may withhold payment to the Contractor of the amount in question until the SO is so satisfied; and

(b) in the case of a nominated sub-contractor or supplier, if the SO certifies that the amount in question has not been paid, the Authority may pay to the sub-contractor or supplier the whole or part of any such amount which shall thereupon be immediately recoverable by the Authority from the Contractor.

The decision of the SO as to whether any such amount has not been paid and of the Authority as to the sum (if any) to be paid to the nominated sub-contractor or nominated supplier shall be final and conclusive.

(7) In paragraph (6) of this Condition, the expression 'nominated sub-contractor or nominated supplier' means a person with whom the Contractor, in compliance with the nomination of the SO or the Authority, has entered into a contract for the execution of work or the supply of things designated as a 'Prime Cost' or 'PC' item in accordance with Condition 38.

Comment

1. The payment conditions of form GC/Works/1 are in theory the best that appear in any of the standard printed contract conditions. They have however a two-fold hidden hazard:

(a) Owing to poor contract administration, documentation and design detail, there is sometimes a greater than usual tendency on Government contracts to build up a backlog of unidentified and disputatious variation adjustment which has the effect of creating a 'hidden' retention fund often substantially in excess of the total fund of 3% that should in theory be held under the conditions. This problem is further exacerbated by any curtailment imposed on appointed quantity surveyors in the exercise of their professional judgement on rating and claim matters. To avoid the unfavourable cash flow that can so easily stem from this cause it is essential that the contractor approaches the progressive agreement of the final account with extreme vigour.

(b) Unlike the ICE Conditions, in which there is a period of 28 days stipulated within which payment should be made from the time of the contractor's payment application, and the JCT form, where the employer is required to honour an architect's certificate within 14 days, there is *no stipulated payment period laid down in these conditions*.

This makes rational assessment of financing costs an act of guesswork on the contractor's part, for although payment is often promptly made by the PSA, there is no remedy should it not be (other than resort to common law).

2. The key points to be noted in the payment conditions are as follows:

(a)   The retention fund to be held by the Authority will be 3% of the value executed plus 3% of the value of materials on site. In the case therefore of a contract valued at an interim stage of £100,000 with a further £10,000 worth of materials properly on site the retention fund held would be 3% of £100,000 i.e. £3,000 and 3% of £10,000 i.e. £300, totalling £3,300 in all.

(b)   The initiative in setting in train the payment procedures rests with the contractor, since the clause states 'the Contractor may at intervals of not less than one month submit claims for payment of advances on account'. The claims for payment are required to be supported by a detailed valuation prepared on the basis of the rates contained in the bill of quantities, rates deduced therefrom, fair rates when the foregoing are not applicable or on a daywork basis.

(c)   In addition under the terms of sub-clause (5) the Contractor is entitled to payment on account of claims, as will be seen from the reference in the clause to Clauses 9(2) or 53(1) which are the 'claims clauses' under this form of contract.

It is however one thing for the conditions to recognise an entitlement and quite another thing to obtain it. Just as with the ICE and JCT forms, where there is also an entitlement to interim payment in respect of claims obtaining, it invariably proves extremely difficult, and with the PSA often even more difficult than on those other two forms, since for the most part one is confronted by 'public accountability'.

If interim payment of claim is to be obtained, it will normally only be achieved on the basis of contractually well founded interim submissions with financial calculations backed up by appreciable supporting detail and evidence.

(d)   One significant amelioration is the provision that if the contract sum exceeds £100,000 (and today most jobs will exceed that sum) the Contractor may make an application at the end of the second week in each month for an interim advance on account without a detailed substantiating valuation being drawn up. The amount likely to be certified would probably be something less than half the average difference between detailed monthly valuations. This provision, which can substantially assist the Contractor's cash flow, is one of which advantage should always be taken but the words 'the amount of any such interim advance shall be an approximate estimate only and the decision of the SO in regard thereto shall be final and conclusive' are to be noted.

Contractors should appreciate that the final and conclusive nature of this right relates to the 'amount' and not the exercise or otherwise of the entitlement to be paid. The clause uses the words 'shall be paid to the Contractor on his application at the end of the second week in each monthly period' and any attempt to deny the Contractor this entitlement should be countered with the utmost vigour.

(e) It has previously been noted that the clause unfortunately contains no specific indication of the period that shall elapse between the date of payment application, an interim certificate being issued and it subsequently being honoured by payment. Furthermore, examination of the clause covering payment to nominated sub-contractors again gives no comparable indication of the elapsed time between the main contractor receiving payment or a certificate, and the stage when he shall in theory have had to pay his nominated sub-contractors. This inadequacy, whilst of some benefit to the main Contractor, may lead to confusion as to whether or not the $2\frac{1}{2}\%$ cash discount, which is a discount normally given for prompt payment, has been fairly earned. It is surely difficult to see how prompt payment can be defined when no period is in fact stipulated in the contract within which such payment ought properly to be made.

3. Sub-clause (6) contains a requirement that the contractor shall satisfy the SO on request that he has paid his sub-contractors or suppliers those sums included in a previous certificate prior to any further payment being made to him. The sub-clause applies to both domestic and nominated sub-contractors.

From the sub-contractor's standpoint the sub-clause unfortunately lacks teeth. It will be noted that it only comes into play if the request is made by the SO, and in this case the contractor has a mandatory obligation to produce evidence of payment. If the SO does not elect to make a request, matters may drift on for weeks. However, if the SO has involved himself in the situation, the remedies are that no further payment may be made to the main contractor in respect of his domestic sub-contractor until payment has been honoured, whereas in the case of the nominated sub-contractor payment may be made direct and monies that would otherwise have been due be deducted from the payment to the main contractor.

Although an arrangement not officially recognised by the contract, the sensible thing on the part of the sub-contractor would be for him to inform the SO in writing of non-payment, and so make the SO alive to a situation which otherwise might go by default, merely because the SO knew nothing of it.

4. For the purposes of this clause the term 'nominated sub-contractor' will be applied to those items that have been covered by prime cost sums in the contract bills or specifications.

Clause 41(1)—Payment of the First Half of the Retention Monies

Upon the completion of the Works to the satisfaction of the SO the Contractor shall be entitled to be paid the amount which the Authority estimates will represent the Final Sum less one half the amount of the reserve and thereafter the Authority may, if he thinks fit, pay further sums in reduction of the reserve.

Comment

1. In the event of the variation account having been kept up to date and rates progressively agreed, with hopefully no uncontemplated loss and expense having been incurred and being the subject of outstanding claim, the exercise by the Authority of the 50% release of the retention money should result in only some $1\frac{1}{2}\%$ of the contract sum being outstanding soon after the issue of the certificate of completion.

2. Unfortunately, as has already been noted, where jobs have been the subject of substantial variation there is often at the completion stage a considerable volume of unidentified variation work tending to increase artificially the retention fund above its theoretical level. Note however that the clause uses the words 'which the Authority estimates will represent the final sum'. If this power is used with a little intelligence and some judicious prodding on the part of the contractor, the serious effect on cash flow which would otherwise come about can be considerably lessened.

It is quite possible that while the parties may not have fully agreed the variation account, a reasonable assessment of the amount of variation additions might, for example, be in order of £10,000. That figure might then be used to assess the final sum, for purposes of retention release, without necessarily having to wait for the completion of the variation account. Bear in mind, in the unlikely event of an over-payment on the part of the Authority being made, that offset can be made against any other contract with any Government Department to recover the over-payment.

3. It should also be noted that the clause contains a power enabling the Authority, if he thinks fit, to pay further sums in reduction of a reserve after the point at which one half of the retention monies have been released and prior to the time when the final payment would be made. If during the progressive work on the final account additional monies are identified, their prompt payment should be obtained by the contractor formally asking for an additional payment and highlighting the authority given by this Clause for it to be made.

4. In this, as in all the other clauses dealing with payment, it will be noted that there is no time-scale for payment laid down and it merely says 'the Contractor shall be entitled to be paid', with no precise indication as to when.

Clause 41(2)—Final Account

As soon as possible after the completion of the Works to the satisfaction of the SO the Quantity Surveyor shall forward one copy of the Final Account to the Contractor.

Comment

1. The clause makes it clear that the onus of responsibility for the preparation of the final account rests formally on the quantity surveyor named under the contract. The old CC/Works/1 conditions required that the contractor provide the final account, rather as under the ICE Conditions; in the writer's opinion this is much the better arrangement, since the power to push things along then rests with the very person most interested in seeing things resolved in a timely way—namely the contractor who depends on cash as the life-blood of his business.

2. The requirement that the final account be provided as soon as possible after the completion of the works to the satisfaction of the SO appears a little anachronistic, since until the maintenance period has elapsed, and the SO has issued a certificate certifying when the works were in a satisfactory state, the final incidence of variations will still not have been known.

3. Unlike both the ICE and JCT forms of contract, no stipulated period of final measurement is given other than by the nebulous form of words 'as soon as possible', leaving the Contractor in a rather difficult position when wishing to make the point that the production of the final account has been unreasonably delayed, and that such delay, being a breach of the contract, is involving him in damage which ought properly to be recoverable.

4. The preparation of the final account will obviously mean that a great deal of work will have to be carried out in conjunction with the contractor. On a large contract it is sensible and beneficial if the professional quantity surveyor named in the contract and his counterpart employed by the contractor agree to share the burden of variation measurement and split the work load between them, thereafter their respective measurements being exchanged for the purposes of checking.

The extent to which this is encouraged in practice must rest upon the demonstrable competence of the contractor's quantity surveyor.

Condition 41(3)—Payment when the Final Account is Agreed after the End of the Maintenance Period

If after the end of the end of the maintenance period specified in the Abstract of Particulars the SO has certified that the Works are in a satisfactory state, and the Final Sum has been calculated and agreed (or in default of agreement has been determined by an Arbitrator appointed under Condition 61) then:

(i)   if the Final Sum exceeds the total amount paid to the Contractor, the excess shall be paid to the Contractor by the Authority; or
(ii)   if the total amount paid to the Contractor exceeds the Final Sum, the excess shall be paid to the Authority by the Contractor.

Comment

1. When the final account has been agreed after the end of the maintenance period, the clause presupposes that any outstanding balance comprising both the second half of the retention monies and any further sums attributable to variation works be paid off in one lump.

The clause caters for both the situation in which the contractor might owe the authority money due to an earlier over-payment and the more usual situation in which the Authority owes the contractor money.

2. If it is apparent that the final account is unlikely to be agreed until after the end of the maintenance period, the Contractor should always endeavour to obtain further payments as previously noted under the terms of Clause 41(1).

Clause 41(4)—Payment when the Final Account has been Agreed before the End of the Maintenance Period

If the Final Sum has been calculated and agreed before the end of the said maintenance period, then:

(i)   if the balance of that sum due to the Contractor exceeds any reserve which the Authority is for the time being entitled to retain, that excess shall be paid to the Contractor by the Authority; or
(ii)   if the total amount paid to the Contractor exceeds the Final Sum, the excess shall be paid by the Contractor to the Authority.

Comment

1. Whilst it is obviously desirable that the final account be agreed as quickly as possible, it appears a little foolhardy that agreement should have been given before the end of the Maintenance Period, since, as has been noted, variations will often be incorporated in the schedule of defects and it would appear more appropriate that the contractor should agree the final account subject to the addition of any items subsequently identified on the schedule of defects and agreed between the parties as being a variation.

2. Again where the final account, agreed subject to the addition of variations relating to defects, identifies monies beyond those entitled to be held at the rate of $1\frac{1}{2}\%$ of the value of the work executed (i.e. the second half of the retention monies), such monies should be paid to the contractor as soon as possible after their identification.

3. The final certificate when issued would then comprise the second half of the retention monies plus any amounts due in respect of variations carried out under the guise of a defect.

4. Provision is once more made for the possibility of balances being due either from the contractor to the Authority or from the Authority to the contractor.

5. It may be useful in arguing the case for payment being made right up to the hilt to emphasise the underlying philosophy of the contract, which supposes that in its concluding stages it is quite possible that payment may have been made so close up to and possibly beyond the final sum that arrangements for a payment requirement from the contractor to the Authority are a perfectly normal and recognised feature of the contract.

Clause 42(1)—Issue of Certificates

The SO shall from time to time certify the sums to which the Contractor is entitled under Conditions 40 and 41. The SO shall also certify the date on which the Works are completed to his satisfaction and after the end of the said maintenance period he shall issue a certificate when the Works are in a satisfactory state.

Comment

1. The certificates required under the contract are:
(a) interim payments certificates issued under Clause 40(3);
(b) the final certificate;
(c) the certificate certifying when the works are completed;
(d) the certificate certifying when the works are in a satisfactory state, i.e. after defects have been made good.

2. The issue of the completion certificate is of considerable importance to the Contractor, since it is from the date so certified that:
(a) liquidated damages liability would be defined (if any);
(b) the first half of the retention monies would be released;
(c) the defects liability period would start to run.

3. The use of the word 'shall' in the clause indicates that there is a mandatory obligation on the SO to issue the various certificates required under the contract. Were the SO to refuse to issue a certificate called for under the contract and as a result the contractor were to suffer loss, such loss would be recoverable stemming from a breach of the Contract.

Clause 42(2)—Interim Payment Certificates

Any interim certificate relating to payment for work done or things for incorporation delivered may be modified or corrected by any subsequent Interim Certificate or by the Final Certificate for payment, and no Interim Certificate of the SO shall of itself be conclusive evidence that any work or thing to which it relates are in accordance with the Contract.

Comment

1. This clause emphasises that an interim certificate exists purely to enable cash on account to be passed to the contractor during the course of construction and to enable the financing of the job to be adequately covered. It in no way validates the value of the work carried out or provides evidence of the satisfactory execution of the work included in the certificate.

2. In practice the work in each certificate will always be the subject of total re-appraisal, since in each interim certificate the works will be totally valued and the amount previously included in any previous payment deducted from the total. In this way it will be appreciated that the only certificate which seeks formally to define the value of the work carried out and to provide any conclusive evidence of its satisfactory execution is the final certificate certifying final payment.

3. Although it is customary for financial auditors to regard the certificates issued as being conclusive evidence of the value of turnover on a particular contract, it must be appreciated that this is, in reality, a practice that is actually unwarranted by the terms not only of these conditions but of both the JCT and ICE Conditions as well.

Clause 42(3)—The Contractor's Right to Question the Certificate

Any dispute as to the Contractor's right to a certificate or as to the sums to be certified from time to time, shall be referred to the Authority whose decision shall be final and conclusive:

Provided that this paragraph shall not apply to a dispute:

(i)   as to a matter in respect of which Condition 40 provides for the decision of the SO to be final and conclusive;
(ii) as to the Contractor's right to a certificate regarding the satisfactory state of the Works after the end of the maintenance period; or
(iii) as to the amount of the balance of the Final Sum due to the Contractor.

Comment

1. Although stated in a convoluted way the point of final decision under the Contract in respect of various certificates is as follows:

(a) as to an interim payment certificate, the SO;
(b) as to whether payments have been made to a nominated sub-contractor, the SO;
(c) as to whether a completion certificate should be issued, the Authority named in the abstract of particulars;
(d) as to the contractor's right to a certificate certifying the satisfactory state of the works at the end of the defects laibility period, an arbitrator;
(e) as to the amount of a final certificate, an arbitrator.

2. The only matters under these conditions that can therefore be referred to an arbitrator will concern the issue of a certificate certifying the satisfactory state of the work, the amount of the final certificate, and such other matters as have not in this clause already been stated as being a matter for final and conclusive decision by the SO or the Authority.

STANDARD LETTERS

Reference: **GC/39**

To: The SO

Dear Sir

Claim for Monthly Payment on Account

We enclose our claim for payment of advance on account of work done and things for incorporation delivered pursuant to Clause 40(3) of the Conditions.

This claim is supported by the appended valuation of work done and items delivered for incorporation.

Yours faithfully

Contractor Limited

Reference: **GC/40**

To: The SO

Dear Sir

Claim for Interim Payment Between Monthly Claims

Since the contract sum for the above project exceeds £100,000 we make application at the end of the second week following the previous monthly claim for the month of .... for an interim advance on account of further work done and things for incorporation since the last valuation pursuant to Clause 40(3) of the Conditions.

The amount of such interim advance being an approximate estimate only we suggest be based on the sum of the two preceding monthly certificates divided by four i.e.

£

Certificate No...
Certificate No...

_____
4 |
_____

_____   =approximate estimate only for
            interim advance

Yours faithfully

Contractor Limited

Reference: **GC/41**

To: The SO

Dear Sir

Submission of Proof of Payment to a Nominated Sub-Contractor or Nominated Supplier

We refer to your request for proof of payment to nominated sub-contractors and suppliers pursuant to Clause 40(6) of the Conditions and now enclose [..............................] in respect of the undermentioned firms:

  (i) ....
  (ii) ....
 (iii) ....
 (iv) ....

Yours faithfully

Contractor Limited

Reference: **GC/42**

To: The SO

Dear Sir

Application for Payment of the First Half of the Retention Money

We refer to the Certificate of Completion of the works dated
.... issued under Clause 42(1) and note that
we may now expect to receive certification and payment of the first half of
the retention monies pursuant to Clause 41(1) of the Conditions.

The amount due for certification we understand to be £............

Yours faithfully

Contractor Limited

Reference: **GC/43**

To: The SO

Dear Sir

*Application for Further Payment after Release of the First Half of the
Retention Money*

Subsequent to the issue of the certificate releasing the first half of the
retention monies, continuing work on the final account/agreement of the
final sum before the end of the period of maintenance has indicated
additional monies to be due to us.

A further interim certificate is therefore requested pursuant to Clause
41(1)/41(4)(i) of the Conditions.

Yours faithfully

Contractor Limited

Reference: **GC/44**

To: The Authority

Dear Sir

Application for Final Payment

We note that the issue of a certificate of satisfactory state of works under Clause 42(1) has been made and that the final sum has been calculated and agreed, leading us to expect the issue of a certificate for the second half of the retention monies and such other balances as may have been identified as outstanding pursuant to Clause 41(3) of the conditions.

The amount due we understand to be £..................

Yours faithfully

Contractor Limited

Reference: **GC/45**

To: The SO

Dear Sir

Request for a Certificate of Completion

We herewith apply for a certificate to be issued certifying the date on which the works were completed to the satisfaction of the SO pursuant to Clause 42(1) of the conditions.

We further undertake to finish any outstanding work of a minor nature, not being such as would prevent the works being used for the purposes intended, during the maintenance period.

Yours faithfully

Contractor Limited

Reference: **GC/46**

To: The SO

Dear Sir

Request for a Certificate of Satisfactory State

Believing all defects notified in the schedule of defects to have been made good, we request a certificate of satisfactory state stipulating the day on which such making good was completed pursuant to Clause 42(1) of the conditions.

Yours faithfully

Contractor Limited

Reference: **GC/SO/47**

To: The Contractor

Dear Sir

Interim/Final Payment Certificate

We acknowledge delivery of your [monthly statement for the month of .... on ....] [statement of final account and supporting documentation on ....] and now issue our Interim/Final Certificate in the cumulative amount of £............................. which after adjustment for retention and previous payment certifies £............................. as due for payment pursuant to Clause 40(3)/stating the amount of .... in our opinion finally due under the contract up to the date of the end of the maintenance period specified which after adjustment for previous payments certifies £............................. as due for payment pursuant to Clause 41(3) of the Conditions.

A copy of the aforementioned document has been sent to the Authority.

Yours faithfully

Superintending Officer

Reference:**GC/SO/48**

To: The Contractor

Dear Sir

Interim Payment Between Monthly Claims

We acknowledge delivery of your interim application for the two-week period following the previous monthly claim for the month of .... on .... and now issue our Interim Certificate in the cumulative amount of £............................. which after adjustment for reserves and previous payment certifies £............................. as due for payment pursuant to Clause 40(3) of the conditions.

A copy of the aforementioned document has been sent to the Authority.

Yours faithfully

Superintending Officer

Reference: **GC/SO/49**

To: The Contractor

Dear Sir

Proof of Payment to Nominated Sub-Contractor

We refer to the nominated sub-contract for ....
.... being carried out by
.... and request proof of payment in respect of the
monies included for them under our certificate no. .... dated
....

Such proof shall be in the form of .... and is
requested pursuant to Clause 40(6)(b) of the conditions.

Yours faithfully

Superintending Officer

Reference: **GC/SO/50**

To: The Contractor

Dear Sir

Final Account

We enclose herewith one copy of the Final Account pursuant to Clause
41(2) of the conditions.

Kindly sign the acknowledgement of receipt, copy attached, and return as
soon as possible.

Yours faithfully

Quantity Surveyor

Reference: **GC/SO/51**

To: The Contractor

Dear Sir

Certificate of Completion

We refer to your letter reference .... dated
.... indicating that you consider the whole of the
works to have been substantially completed [and to have passed such
final tests as are prescribed in the Contract].

[We note your undertaking to finish any minor outstanding work during
the Maintenance Period.]

We agree the whole of the works to be substantially complete and hereby
issue our Certificate of Completion, stating the date of completion as
being .... pursuant to Clause 42(1) of the
conditions.

Yours faithfully

Superintending Officer

Reference: **GC/SO/52**

To: The Contractor

Dear Sir

Certificate of Satisfactory State

We enclose herewith our Certificate of Satisfactory State stating the date
on which the contractor completed his obligations to complete and

maintain the works to our satisfaction pursuant to Clause 42(1) of the conditions.

Yours faithfully

Superintending Officer

CHAPTER 16

Remedies and Powers of Determination

Clause 43—Recovery of Sums Due from the Contractor

Whenever under the Contract any sum of money shall be recoverable from or payable by the Contractor such sum may be deducted from or reduced by the amount of any sum or sums then due or which at any time thereafter may become due to the Contractor under or in respect of the Contract or any other contract with the Authority or with any Department or Office of Her Majesty's Government.

Comment

1. This clause constrains the contractor's freedom of commercial action under the contract, since he must always have at the back of his mind this overriding right of offset against any other of his contracts undertaken with a Government body. This clause combined with the fact that there is no contractual right on the contractor's part to determine the contract creates a very one-sided contract as regards remedies.

Remember that of all the forms of contract used in the construction industry in the UK, this is the only one widely employed and drafted unilaterally by the employer.

2. The circumstances in which the Authority could exercise its powers under this clause might be as follows:

  (a) levying additional costs arising from a purported wrongful determination of the contract by the Contractor;
  (b) levying liquidated damages against the contractor where the amount of damages is greater than the retention fund held under the contract;
  (c) where the value of works have been found to be over-certified and repayment is required.

Clause 44—Special Power of Determination by the Authority

(1) The Authority shall, in addition to any other power enabling him to determine the Contract, have power to determine the Contract at any time by notice to the Contractor, and upon receipt by the Contractor of the notice the Contract shall be determined but without prejudice to the rights of the parties accrued to the date of determination and to the operation of the following provisions of this Condition.

(2) (a) The Authority shall as soon as practicable, and in any case not later than the expiration of 3 months from the date of such notice or of the period up to the date for completion, whichever is the shorter, give directions (with which the Contractor shall comply with all reasonable despatch) as to all or of any of the following matters:

   (i)   the performance of further work in accordance with the provisions of the Contract;
   (ii)   the protection of work executed under the Contract in compliance with directions given under sub-paragraph (i) above;
   (iii)   the removal from the Site of all things whether or not they were for incorporation;
   (iv)   the removal of any debris or rubbish and the clearing and making good of the Site;
   (v)   the termination or transfer of any sub-contracts and contracts (including those for the hire of plant, services and insurance) entered into by the Contractor for the purposes of or in connection with the Contract; or
   (vi)   any other matter arising out of the Contract with regard to which the Authority (whose decision on the matter shall be final and conclusive) decides that directions are necessary or expedient.

   (b) The Authority may at any time within the period referred to in sub-paragraph (a) above by notice to the Contractor vary any directions so given or give fresh directions as to all or any of the matters specified in that sub-paragraph.

(3) (a) In the event of the determination of the Contract under this Condition there shall be paid to the Contractor:

   (i)   the net amount due, ascertained in the same manner as alterations, additions and omissions under the Contract, in respect of work executed in accordance with the Contract up to the date of determination;
   (ii)   the net amount due, ascertained in the same manner, in respect of works or services executed in compliance with directions given by the Authority under paragraph 2(a)(i), (ii), (iii), (except in so far as it relates to things which were for incorporation being things which the Contractor elects to retain), (iv) and (vi) of this Condition;
   (iii)   the net amount due on the basis of fair and reasonable prices for any things for incorporation which the Contractor, with the consent of the Authority, has elected not to retain as his in own property and which at the date of determination:

      (*a*)   had been supplied by the Contractor and properly brought on the Site by him and at his expense in connection with and for the purpose of the Contract, that had not been incorporated in the Works;
      (*b*)   were in course of manufacture by the Contractor in connection with and for the purposes of the Contract and were not lost or damaged by reason of any of the accepted risks; and

(iv)   any sum expended by the Contractor on account of the determination of the Contract in respect of the uncompleted portion of any sub-contract and contract (including those for the hire of plant, services and insurances) entered into by the Contractor for the purposes of or in connection with the Contract, to the extent to which it is reasonable and proper that the Authority should reimburse that sum; and

(v)   any sum expended by the Contractor in respect of any contract of employment which is expended on account of the determination of the contract or which, but for this provision, would represent an unavoidable loss by reason of the determination, to the extent to which it is reasonable and proper that the Authority should reimburse that sum.

(b) If the Works or any part thereof or any things to which sub-paragraph (a)(iii)(*a*) above relates are at the date of determination, or if directions are given in pursuance of paragraph (2)(a) of this Condition at the date for completion of the Works, lost or damaged by reason of any of the accepted risks and such loss or damage was not occasioned by any failure on the part of the Contractor to perform his obligations under Condition 25, the net amount due shall be ascertained as if no loss or damage had occurred.

(c) There shall be deducted from any sum payable to the Contractor under this paragraph the amount of all payments previously made to the Contractor in respect of the Contract, and the Authority shall have the right to retain any reserves accumulated in his possession at the date of determination until the final settlement of all claims made by the Contractor under this paragraph.

(d) The Contractor shall for the purposes of this paragraph keep such wage books, time sheets, books of account and other documents as are necessary to ascertain the sums payable hereunder and shall at the request of the Authority provide (verified in such manner as he may require) any documents so kept and such other information as he may reasonably require in connection with matters arising out of this Condition.

(4) All things not for incorporation which are brought on the Site at the Contractor's expense shall (whether damaged or not) re-vest in and be removed by him as and when they cease to be required in connection with the directions given by the Authority under paragraph (2)(a)(i), (ii), (iii), (iv) and (vi) of this Condition. The Authority shall be under no liability to the Contractor in respect of the loss thereof or damage thereto caused by reason of any of the accepted risks.

(5) If upon the determination of the Contract under this Condition the Contractor is of the opinion that he has suffered hardship by reason of the operation of this Condition he may refer the circumstances to the Authority, who, on being satisfied that such hardship exists, or has existed, shall make such allowance, if any, as in his opinion is reasonable, and his decision on that matter shall be final and conclusive.

(6) The Contractor shall, in any substantial sub-contract or contract made by him in connection with or for the purposes of the Contract take power to determine such sub-contract or contract in the event of the determination of the Contract by the Authority upon terms similar to the terms of this Condition, save

that the name of the Contractor shall be substituted for the Authority throughout except in paragraphs (3)(a)(iii), (3)(d) and (5).

Comment

1. The nature of this clause in a standard construction contract is unique (although similar terms may be found in contracts used in the petro-chemical industry). It allows the Authority to determine a contract at will without being guilty of breach and with no underlying failure, fault or breach on the part of the contractor.

The one-sided nature of the clause is then compounded, as under sub-clause (5) the decision of the Authority concerning financial hardship suffered by the contractor is final and conclusive, thereby curtailing the normal arbitration rights.

2. This unusual clause has obviously been incorporated to cater for those situations in which there is a change of Government policy. Whilst one can not cavil at the contract having legislated for such a probability, one can hardly be enthusiastic at the one-sided manner in which the consequences of such determination are to be resolved.

3. The 'as of right' reimbursement to the contractor is limited within sub-clause (3)(a) purely to the direct site costs valued at the contract rates or at rates analogous thereto, reimbursement for materials properly on site, and sub-contract costs where the Contractor may have financial liability, again however limited to those direct costs immediately associated with the work.

There is no as-of-right basis of recovery for loss of profit and overhead cover on any section of the work not carried out and which in a normal breach of contract would rate as quite straightforward reimbursable damage.

4. The only way open to the contractor to seek recovery for damage beyond the direct site costs previously detailed will be by resort to sub-clause (5). Thus if the contractor believes he has suffered hardship, a probability bordering on the inevitable when a contract has been determined, he must rely on the good offices of the Authority to reimburse him in a situation in which the Authority is acting as both judge and jury in its own case.

5. Unfair as this may appear to the contractor, it is essential under sub-clause (6) that he incorporates a comparable term in any substantial sub-contract he enters into in order that the right of recovery of the sub-contractor against himself in the event of such determination be restricted to the limited scope of recovery under the terms of the main contract conditions available to the main contractor.

Should this condition not have been incorporated in a sub-contract, it is quite possible that the sub-contractor could obtain an appreciably higher level of recovery against the main contractor than the contractor would subsequently be able to recover from the Authority. If this occured, little sympathy could be expected from the Authority if the contractor had not protected himself under the terms of his sub-contract as specifically required by sub-clause (6), since by not having done so he would have been in breach of his contract with the Authority.

Clause 45—Determination of Contract Due to Default or Failure of the Contractor

The Authority may without prejudice to the provisions contained in Condition 46 and without prejudice to his rights against the Contractor in respect of any delay or inferior workmanship or otherwise, or to any claim for damage in respect of any breaches of the Contract and whether the date for completion has or has not elapsed, by notice absolutely determine the Contract in any of the following cases, additional to those mentioned in Condition 55 hereof:

(a)   if the Contractor, having been given by the SO a notice to rectify, reconstruct or replace any defective work or a notice that the work is being performed in an inefficient or otherwise improper manner, shall fail to comply with the requirements of such notice within seven days of the service thereof, or if the Contractor shall delay or suspend the execution of the Works so that either in the judgement of the SO he will be unable to secure completion of the Works by the date for completion or he has already failed to complete the works by that date;

(b)   (i)   if the Contractor, being an individual, or where the Contractor is a firm, any partner in that firm, should at any time become bankrupt, or shall have a receiving order or administration order made against him or shall make any composition or arrangement with or for the benefit of his creditors or shall make any conveyance or assignment for the benefit of his creditors, or shall purport to do so, or if in Scotland he shall be insolvent or notour bankrupt, or any application shall be made under any Bankruptcy Act for the time being in force for sequestration of his estate, or a trust deed shall be granted by him for behoof of his creditors; or

(ii)   if the Contractor, being a company, shall pass a resolution, or if the Court shall make an order, that the company shall be wound up, or if the Contractor shall make an arrangement with his creditors or if a receiver or manager on behalf of a creditor shall be appointed, or if circumstances shall arise which entitle the Court or a creditor to appoint a receiver or manager or which entitle the Court to make a winding up order; or

(c)   in the case where the Contractor has failed to comply with Condition 56 if the Authority (whose decision on this matter shall be final and conclusive) shall decide that such failure is prejudicial to the interests of the State:

Provided that such determination shall not prejudice or affect any right of action or remedy which shall have accrued or shall accrue thereafter to the Authority.

Comment

1. This Clause deals with the more usual circumstances that would enable an employer to determine a contract and similar provisions will be found in most of the standard forms.

Unlike the JCT form, however, there is no reciprocal clause that deals with those circumstances in which the contractor might be entitled to determine the contract. This appears to stem from a belief underlying these conditions that a Government Department could not be a party to error or omission that would warrant determination. Unfortunately events do not always bear out this proposition.

2. The grounds set down for determination by the Authority are then as follows:

  (a) Where the Contractor has received and has failed to comply within seven days with a notice relating to:

      (i)   rectification, reconstruction or replacement of any defective works;

      (ii)   carrying out of work in an inefficient or improper manner;

      (iii) delaying or suspending the execution of the work so as to jeopardise the completion of the works within the stipulated period.

  (b) Bankruptcy of the contractor.

  (c) An act of corruption pursuant to Clause 55.

  (d) Failure by the contractor to comply with the requirements of the Authority in regard to the admittance to the site of individuals pursuant to Clause 56.

Clause 46—Settlement in the Event of Determination Pursuant to Condition 45

If the Authority, in the exercise of the power contained in Conditions 45 or 55, shall determine the Contract, the following provision shall take effect.

Comment

1. A précis of the respective rights and liabilities upon determination is as follows:

  (a) Monies due from the Authority to the contractor shall cease to be due.

  (b) The Authority may employ another contractor to complete the works, who may enter upon the site and take possession of all materials, goods, and plant thereon.

(c) Other than in the case of bankruptcy, any sub-contract shall be assigned to the Authority. The balance of the price due to the sub-contractor shall then be paid by the Authority. If amounts have previously been certified by the Authority for payment to the sub-contractor and these have not been so paid, then such amount may forthwith be recoverable by the Authority from the contractor.

(d) The Authority may pay to any nominated sub-contractor or nominated supplier sums due to them that previously had been certified by the SO but which had not been paid by the contractor. Such amounts may again be recovered by the Authority from the contractor.

(e) The SO shall certify the cost of completing the works, which is deemed to include:

    (i)   the cost of labour and materials together with a reasonable allowance for superintendance and establishment charges in completing the works;

    (ii)   the cost of work executed by other contractors;

    (iii)   the amount of liquidated damages which might accrue.

2. The clause also covers the somewhat academic situation in which the cost of completing the works as previously outlined would be less than if the works had been finished in total by the original contractor. The chance of such a situation occurring appears so remote that we will not consider the matter.

3. In the more likely case where the completion cost is greater than it would have been had the contractor completed the whole of the original works, then the Authority may sell any plant or materials remaining on the site and set the proceeds against the outstanding deficit. If the deficit is still not covered, the remaining sum may be recovered by offset against other contracts as previously indicated pursuant to Clause 43.

STANDARD LETTERS

Reference: **GC/SO/53**

To: The Contractor

Dear Sir

Notification of Special Power of Determination by the Authority

We hereby give notice and invoke the special power of determination pursuant to Clause 44 of the conditions and determine the contract as of
....

In accordance with the provision of Clause 44(2) you are requested to comply with the following:

....
....
....
....
....

Any sum due to the contractor shall be ascertained under Clause 44(3)(a).

Yours faithfully

Superintending Officer

Reference: **GC/SO/54**

To: The Contractor

Dear Sir

Determination of Contract due to Default of the Contractor

We hereby give notice and absolutely determine your contract pursuant to Clause 45 on the ground of:-

( )   failure after seven days notice to rectify, reconstruct or replace defective work/cease carrying out the work in an inefficient or improper manner/cease delaying or impeding the work so as to jeopardise completion within the stipulated period

( )   bankruptcy

( )   an act of corruption

( )   failing to meet the requirements of the Authority in regard to the admittance to the site of certain persons.

Yours faithfully

Authority

CHAPTER 17

Third Party Liabilities, Damage to Highways and Emergency Powers

Clause 47(1) and (2)—Indemnity Against Third Party Liability

(1) This Condition applies to any personal injury or loss of property (not being a loss of property to which Condition 26 applies) which arises out of or in any way in connection with the execution or purported execution of the Contract.

(2) Subject to the following provisions of this Condition, the Contractor shall:

(a) be responsible for and reinstate and make good to the satisfaction of the Authority, or make compensation for, any loss of property suffered by the Crown to which this Condition applies;

(b) indemnify the Crown and servants of the Crown against all claims and proceedings made or brought against the Crown or servants of the Crown in respect of any personal injury or loss of property to which this Condition applies and against all costs and expenses reasonably incurred in connection therewith;

(c) indemnify the Crown against any payment by the Crown in order to indemnify in whole or in part a servant of the Crown against any such claim, proceedings, costs or expenses; and

(d) indemnify the Crown against any payment by the Crown to a Crown servant in respect of loss of property to which this Condition applies suffered by that servant of the Crown and against any payment made under any Government provision in connection with any personal injury to which this Condition applies suffered by any servant of the Crown.

Comment

1. This condition is the normal type of third-party indemnity that will be found in any of the standard contract forms. Reference to the exclusion of a loss to which Clause 26 applies refers to loss or damage to the works themselves or to materials properly on site for incorporation in those works. Such matters are of course not a third-party liability between the parties.

2. The third-party liability under the clause is extremely widely drafted and covers:

(a) damage to Government property other than the actual works,

(b) personal injury to a third party,

(c) injury to the property of a third party,

(d) indemnifying a civil servant in the event of a third party liability claim being brought against him, and,

174

(e) liability to compensate the Crown in the event of a civil servant suffering injury and obtaining compensation from the Crown.

Clause 47(3)—Contributory Liability of the Authority

If the Contractor shows that any personal injury or loss of property to which this Condition applies was not caused nor contributed to by his neglect or default or that of his servants, agents or sub-contractors, or by any circumstances within his or their control, he shall be under no liability under this Condition, and if he shows that the neglect or default of any other person (not being his servant, agent or sub-contractor) was in part responsible for any personal injury or loss of property to which this Condition applies, the Contractor's liability under this Condition shall not extend to the share in the responsibility attributable to the neglect or default of that person.

Comment

1. Having in the first part of the clause indicated the extremely comprehensive nature of the contractor's third-party liability, this sub-clause provides the contractor with some means of escape from its all-pervading nature.

2. In seeking to mitigate his responsibilities under the condition, the onus rests upon the contractor to demonstrate that the injury or loss complained of is not totally or is only partially his fault.

3. The grounds that would give rise to this total or partial offset of financial responsibility are as follows:

(a) that it was neither caused nor contributed to by the neglect of the Contractor;

(b) that it was neither caused nor contributed to by one of his servants, agents or sub-contractors,

(c) that it was neither caused nor contributed to by any circumstances that was within the contractor's control or the control of his servants, agents or sub-contractors;

(d) that another party not being his servant, agent or sub-contractor has caused the loss, either in whole or in part. To the extent that the loss is attributable to a third party, the loss will be apportioned between the contractor and the Authority and presumably the Authority would then have to recover such part as might be appropriate from the third party.

Clause 47(4)—Procedure Relating to Resolution of Matters of Third-Party Liability

(a) The Authority shall notify the Contractor of any claim or proceeding made or brought in respect of any personal injury or loss of property to which this Condition applies.

(b) If the Contractor admits that he is liable wholly to indemnify the Crown in respect of such claim or proceeding, and the claim or proceeding is not an excepted claim, he, or, if he so desires, his insurers, shall be responsible (subject to the condition imposed by the following sub-paragraph) for dealing with or settling that claim or proceeding.

(c) If in connection with any such claim or proceeding with which the Contractor or his insurers are dealing, any matter or issue shall arise which involves or may involve any privilege or special right of the Crown (including any privilege or right in relation to the discovery or production of documents) the Contractor or his insurers shall before taking any action thereon, consult the legal adviser to the Authority and act in relation thereto as may be required by the Authority, and if either the Contractor or his insurers fail to comply with this sub-paragraph, sub-paragraph (b) above shall cease to apply.

(d) For the purposes of this paragraph 'an excepted claim' means a claim or proceeding in respect of a matter falling to be dealt with under a Government provision, or a claim or proceeding made or brought by or against a servant of the Crown.

Comment

The procedure laid down in this sub-clause pre-supposes that:

(a) the Authority notifies the contractor of a claim;

(b) if the contractor admits liability for the claim, he or his insurers become solely responsible for settling that claim;

(c) where any matter of privilege or special right affecting the Crown may be involved the contractor is required to be advised by the legal advisor to the Authority and shall abide by the advice given.

Clause 47(5)—Limitation of Liability

Where any such claim or proceeding as is mentioned in paragraph (2)(b) or (c) of this Condition is settled otherwise than by the Contractor or his insurers, he shall not be required to pay by way of indemnity any sum greater than that which would be reasonably payable in settlement having regard to the circumstances of the case and in particular to the damages which might be recoverable at law.

Comment

1. This limitation clause applies to a situation in which the Authority might themselves settle and pay a third-party claim, and where the amount paid in compensation, possibly because of some policy decisions, was rather more than would be strictly enforceable at law; the contractor would not then be liable in respect of the amount of the excess sum paid.

2. The clause then goes on to define more fully certain of the terms used elsewhere in the Condition and in particular the following should be noted:

(a) 'Loss of property' includes damage to property, loss of profits and loss of use. Thus for example if the contractor were to damage the manufacturing premises of a third party, in a way for which he was liable, to such an extent as to cause a manufacturing process to cease for a period of time, he would be liable not only for the damage to the property itself but also the loss of profits that accrued as a natural and foreseeable consequence.

(b) 'Personal injury' includes sickness and death.

(c) 'Servant of the Crown' applies to both persons who are servants of the Crown at the time of the personal injury or loss of property and to persons who subsequently cease to be servants of the Crown.

(d) 'Government provision' means any regulation applicable to a servant of the Crown that makes provision for continuance of pay for sick pay, or allowance to families of the person concerned, or payments made in respect of disablement or injury.

Clause 48—Damage to Public Roads

(1) Notwithstanding the provisions of Condition 47, the Authority shall, subject to the following provisions of this Condition, indemnify the Contractor against all claims and proceedings made or brought against the Contractor in respect of any damage to highways, roads or bridges communicating with or on the routes to the Site, including any mains, pipes or cables under such highways, roads or bridges, caused by any extraordinary traffic of the Contractor or sub-contractor or supplier in connection with the Works.

(2) The Contractor shall take all reasonable steps to prevent such highways, roads and bridges from being subjected to damage by extraordinary traffic as aforesaid and in particular but without prejudice to the generality of the foregoing shall select routes, choose and use vehicles and restrict and distribute loads so that any such extraordinary traffic shall be limited so far as is reasonably practicable.

(3) The Contractor shall, without prejudice to his obligations under paragraph (2) of this Condition, comply with such instructions regarding any of the matters mentioned in the said paragraph as may be given to him from time to time in writing by the SO.

(4) The Contractor shall notify the Authority of any claim or proceeding made or brought in respect of any damage to highways, roads or bridges by extraordinary traffic to which this Condition applies and thereafter the Authority shall be responsible for dealing with or settling that claim or proceeding:

Provided always that if it is decided by the Authority that any such claim or proceeding is due, wholly or in part, to any failure by the Contractor to comply with the provisions of paragraph (2) of this Condition or with the SO's instructions under paragraph (3) thereof, then the Authority may recover from the Contractor so much of the costs and expenses incurred by the Authority in connection with such claim or proceeding, as is due to the failure of the Contractor in that respect.

(5) 'Extraordinary traffic' for the purpose of this Condition means extraordinary traffic to which Section 62 of the Highways Act 1959 (or in relation to Scotland, Section 54 of the Road Traffic Act 1930) or any statutory modification or re-enactment thereof as for the time being in force, applies.

Comment

1. This clause is reminiscent of the comparable Clause 30 of the ICE Conditions of Contract, whilst being rather more simply drafted and avoiding the complication of endeavouring to differentiate between traffic bearing goods and materials for permanent works, temporary works and construction plant or carriage undertaken by a specialist haulier. Under these conditions, whatever is being hauled to the site, providing it is for the purpose of the works, falls within the orbit of the clause.

2. It should be appreciated that the clause applies to all highways communicating with the site whether they be public or private. It is unfortunate that the clause bears a heading 'damage to public roads', since by superficial examination this would appear to imply that the clause does not apply to private roadways. In the context of Government work this is a very important consideration, since commonly work will be undertaken on a site in a substantial area of Government-owned properties served by Government-owned roads.

If the SO or Authority at any time seek to argue that the contractor is responsible for damage to Government private roads, reference should be made to Clause 1(5) in which it states 'the headings of these Conditions shall not affect the interpretation thereof'. Returning then to the wording of Clause 48, we see the phrase used 'any damage to highways, roads or bridges communicating with or on the routes to the site'. The clause does not differentiate as to whether the road is a public or private road and the contractor should always argue that the protection afforded by this clause extends to any highway of whatever type.

3. Providing the contractor has taken all reasonable steps to prevent damage, as detailed in sub-clause (2), and has complied with any instructions issued by the SO under sub-clause (3), the Authority will indemnify the contractor against any damage caused.

4. However, to the extent that the contractor has not taken reasonable measures to avoid damage or complied with the instructions of the SO in regard thereto, he will lose the benefit of the indemnity offered.

Clearly, however, much as this situation may appear black and white in theory, there is always liable to be argument as to what was reasonable or unreasonable in the measures that the contractor took or failed to take.

5. Where the SO does issue instructions concerning prevention of damage to the highway, pursuant to his powers of instruction bestowed under Condition 7(1)(m), causing the contractor expense beyond that provided for in or reasonably contemplated by the contract, then he may claim for such additional expense under Condition 9(2).

Clause 49—Emergency Powers

If in the opinion of the SO any urgent measures shall become necessary during the progress of the execution of the Works to obviate any risk of accident or failure, or if, by reason of the happening of any accident or failure or other event in connection with the execution of the Works, any remedial or other work or repairs shall become urgently necessary for security and the Contractor be unable or unwilling at once to carry out such measures or work or repair, the Authority may by his own or other workpeople carry out such measures or execute such work or repair as the SO may consider necessary. If the measures carried out or the work or repair so executed by the Authority shall be such as the Contractor is liable under the Contract to carry out or execute at his own expense, all costs and expenses so incurred by the Authority shall be recoverable from the Contractor.

Comment

1. This condition is very similar to the comparable Clause 62 of the ICE Conditions.

2. The underlying philosophy behind the contract is that remedial works shall be carried out against the schedule of defects and that these will enable such work to be executed economically and in reasonable quantity at any particular time.

3. The contract recognises, however, that there may on occasion be an overriding need to have work carried out as a matter of urgency. In this situation the SO shall first require the contractor to carry out the work and in the event of his being unable to do so within a reasonable time, alternative resources may be employed to carry it out and the costs so incurred be deducted from the amounts due to the contractor—always supposing of course that the matter of urgent repair is one for which the contractor would otherwise be responsible under the Contract.

STANDARD LETTERS

Reference: **GC/47**

To: The Authority

Dear Sir

Notification of Contributory Liability of the Authority in Relation to Third Party Liability

We refer to the notification of claim in respect of third-party liability received from the Authority dated .... relating to .... and admit no/only partial liability to indemnify the Crown in relation to the matter the subject of claim.

The personal injury/[and]/loss of property concerned was solely/partially due to the neglect or default of a person not being our servant, agent or sub-contractor, namely .... and no/only partial liability is therefore accepted pursuant to Clause 47(3) of the Conditions.

Yours faithfully

Contractor Limited

Reference: **GC/48**

To: The Authority

Dear Sir

Notification in Respect of Claim for Damage to the Highway/a Bridge

We give notification of a claim being made against us in respect of damage purportedly caused by extraordinary traffic to

.... and direct this matter to your attention pursuant to your responsibilities for dealing with and settling any such claim or proceeding under Clause 48(4) of the Conditions.

The relevant papers are appended.

Yours faithfully

Contractor Limited

Reference: **GC/SO/55**

To: The Contractor

Dear Sir

Notification of Intention to Invoke Emergency Powers

We herewith give notice of an emergency matter in the form of .... calling for urgent measures pursuant to Clause 49 of the Conditions.

Any delay in the execution of the necessary work may result in the work being carried out by the Authority or other workmen and to the extent that the work should be done at your expense under the contract the necessary sum will become due from you/be deducted from monies otherwise due to you.

Yours faithfully

Superintending Officer

CHAPTER 18

Independent Contractors, Prolongation and Disruption Expenses

Clause 50—Facilities for Other Contractors

The Authority shall have power at any time to execute other works (whether or not in connection with the Works) on the Site contemporaneously with the execution of the Works and the Contractor shall give reasonable facilities for such purpose:

Provided that the Contractor shall not be responsible for damage done to such other works except in so far as such damage has been caused by the negligence, omission or default of his workpeople or agents; and any damage done to the Works in the execution of such other works shall, for the purposes of Condition 26(2), be deemed to be damage which is wholly caused by the neglect or default of a servant of the Crown acting in the course of his employment as such.

Comment

1. This clause deals with a situation which is one of the commonest sources of delay and inconvenience to a contractor. For a variety of reasons, some other contractor or duly authorised authority may be required to perform works on or about the site, and the contractor having no privity of contract with such independent contractor will often find himself in an extremely difficult position when it comes to the control and programming of those independent activities which may be crucial to the completion of his Works.

2. The clause does not specifically mention 'statutory authorities' in the context of other works being carried out contemporaneously with the execution of the works, but as the hazard of work by statutory authorities is not mentioned elsewhere in the conditions, the contractor must argue that their presence on site, and the interface problems between their work and his own, fall within the scope of this clause.

3. It will be noted that the clause uses the phrase 'give reasonable facilities for such purpose'. The giving of facilities applies not only to material aid but also the mere allowing of the other organisation to work on or about the contractor's site. Thus if the independent contractor, by his mere presence on site, should delay the contractor, the allowing of the organisation on site and the delay imposed by them is of itself the provision of a facility.

4. The first edition of these conditions published in November 1973 incorporated a claims provision for such additional expense as might be caused actually within the clause. Edition 2 has deleted this facility but has incorporated it in Clause 53(1)(c) in substitution.

5. For the purposes of third-party liability, the work carried out by an independent contractor features as the work and property of the third party. To the extent that the contractor may cause damage to that other work as a result of negligence, omission or default of his own workpeople he, the contractor, will be responsible for such damage as a third-party liability under the provisions of Clause 26(2).

The chain of remedy in this situation is that the contractor indemnifies the Authority against liability for damage to the property of a third party, and would therefore pay the Authority the compensation concerned, whilst the Authority would in its turn pay from the indemnity monies recovered the compensation to the independent contractor.

6. The converse of the arrangement outlined above is that if the independent contractor damages the work of the contractor, the contractor obtains remedy from the Authority, who in its turn seeks compensation under indemnity from the other contractor.

Under no account should the contractor allow himself to be drawn into a situation by the SO, or the Authority, in which he is directed to obtain recovery direct from the independent contractor.

Clause 53—Prolongation and Disruption Expenses

(1) If:

(a)   complying with an SO's instruction;
(b)   the making good of loss or damage falling within Condition 26(2);
(c)   the execution of works pursuant to Condition 50; or
(d)   delay in the provision of any of the items specified in paragraph (2) of this Condition

unavoidably results in the regular progress of the Works or of any part thereof being materially disrupted or prolonged and in consequence of such disruption or prolongation the Contractor properly and directly incurs any expense in performing the Contract which he would not otherwise have incurred and which is beyond that otherwise provided for in or reasonably contemplated by the Contract, the Contract Sum shall, subject to paragraph (3) of this Condition and to Condition 23, be increased by the amount of that expense as ascertained by the Quantity Surveyor:

Provided that there shall be no such increase in respect of expense incurred in consequence of the making good of loss or damage falling within Condition 26(2) except where the Contractor is entitled to payment under that provision, and where his entitlement to payment under that provision is limited to a proportionate sum any such increase in respect of expense so incurred shall be limited in like manner.

(2) The items referred to in sub-paragraph (1)(d) of this Condition are:

(a)   any drawings, schedules, levels or other design information to be provided by the SO and to be prepared otherwise than by the Contractor or any of his sub-contractors;

(b)   any work the execution of which, or thing the supplying of which, is to be undertaken by the Authority or is to be ordered direct by him otherwise than from the Contractor and is to be so undertaken or ordered otherwise than in consequence of any default on the part of the Contractor;

(c)   any direction from the Authority or the SO regarding the nomination or appointment of any person, or any instruction of the SO or consent of the Authority, to be given under Condition 38(4).

(3) It shall be a condition precedent to the Contract Sum being increased under paragraph (1) of this Condition:

(a)   in the case of expense incurred in consequence of an SO's instruction that the instruction shall have been given or confirmed in writing and shall not have been rendered necessary as a result of any default on the part of the Contractor;

(b)   in the case of expense incurred in consequence of delay in the provision of any of the items specified in paragraph (2) of this Condition, that, except where a date for the provision of the relevant item was agreed with the SO, the Contractor shall, neither unreasonably early nor unreasonably late, have given notice to the SO specifying that item and the date by which it was reasonably required; and

(c)   in any case that:

(i)   The Contractor, immediately upon becoming aware that the regular progress of the Works or of any part thereof has been or is likely to be disrupted or prolonged as aforesaid, shall have given notice to the SO specifying the circumstances causing or expected to cause that disruption or prolongation and stating that he is or expects to be entitled to an increase in the Contract Sum under that paragraph;

(ii)   As soon as reasonably practicable after incurring the expense the Contractor shall have provided such documents and information in respect of the expense as he is required to provide under Condition 37(2).

Comment

1. This clause is a matter of radical redrafting and addition to the contract as compared with the arrangements of the earlier first edition. All claims under the earlier edition had to be lodged under Clause 9, in which the contractor's entitlements were far less lucidly stated.

2. The present clause envisages expense beyond that otherwise provided for in or reasonably contemplated by the contract, and arising as a result of the SO's instructions, but not involving prolongation and disruption as being recovered under Condition 9(2), whereas prolongation and disruption expenses, which might or might not be attributable to an SO's instruction, are recoverable under the Clause we are presently considering.

3. It is to be noted that the Clause itself talks of 'disruption or prolongation' indicating the possibility of disruption expense without prolongation, or prolongation expense without disruption or both together. Again the heading to the clause in the document is slightly misleading in that it talks of 'prolongation and disruption expenses' seeming to infer that the two necessarily travel together and that therefore, only if the contractor has been granted an extension of time, is it possible to contemplate disruption expenses. This is not the case.

4. The clause is extremely broadly drafted in the context of those things that it would recognise as possible contributory factors creating disruption or prolongation.

In summary they are:

(a) Complying with any of the instructions issued by the SO, which under the terms of Clause 7 covers:
  (i) variation,
  (ii) discrepancy between the contract documents,
  (iii) removing things from the site and substituting other things therefor,
  (iv) removal and re-execution of work,
  (v) the order in which the work is to be carried out,
  (vi) the hours to be worked and the extent of overtime or nightwork,
  (vii) suspension of work,
  (viii) replacing staff,
  (ix) opening up for instruction,
  (x) making good defects,
  (xi) carrying out emergency work,
  (xii) using materials obtained from excavations,
  (xiii) any other matter considered expedient.
(b) Making good loss or damage to the works.
(c) Provision of services for independent contractors or statutory authorities.
(d) Delay in the provision of construction information, work carried out by other independent contractors, free issue materials, and the late nomination of a nominated sub-contractor or supplier.

The only commonly encountered source of employer default that does not appear to have been overtly included in this range of possible causes of prolongation and disruption is failure to provide site access at the appropriate time. However, even here the ground would appear to be covered, albeit in a slightly disguised way, through the authority of the SO to issue an instruction under Clause 7(1)(g) suspending the execution of the Works or any part thereof. Clearly if the whole or part of the site was not available, the SO would have to issue such a suspension instruction.

5. The right to register claims under Clause 53 is subject to a number of conditions precedent of the type commonly encountered under most other forms of contract in that:

(a) If the claim arises as a result of an SO's instruction, the instruction must have been given in writing or confirmed in writing by the SO.

(b) Where the claim arises as the result of the provision of late construction information, work by independent contractors, free issue material or nomination, the contractor must have made application for such information or supply at a time not unreasonably early or late having regard to the time when the item was required, save only that if the date for the provision had already been agreed with the SO further application would not necessarily have to be made. This would appear to cater for the situation in which an approved construction programme had been prepared and agreed with the SO which iindicated the times on which various items of supply and information were to be provided.

(c) Having ensured that the criteria established by conditions precedent (a) or (b) above had been fulfilled, the contractor has then to give notice immediately upon becoming aware that the regular progress of the works is being disrupted or prolonged. The notice is required to detail the circumstances causing the disruption or prolongation and to state that the contractor expects to be entitled to increase in the contract sum. Thereafter the contractor is required to provide documentary information concerning the matter as soon as reasonably practicable after incurring the expense.

6. As on all the standard forms of contract, the service of notice is absolutely crucial to the maintenance of one's claim rights under the contract. The chances of a flexible attitude being adopted to this sort of thing on a Government-sponsored contract is probably even less than would be experienced with other employers. Unless the notice is served in a timely way, the contractor faces the very real possibility of losing his entitlement to claim in a situation where under every other ground the entitlement is supportable and totally justified.

A contractor cannot afford to hesitate in the serving of notices while agonising over whether or not by the serving of notice he may upset or antagonise the SO. If he wants to stay in business, he has no choice; if occasion demands it, the notice must be served and must be served as soon as possible.

STANDARD LETTERS

Reference: **GC/49**

To: The SO

Dear Sir

Notice of Prolongation and Disruption Expenses Associated with the Works of an Independent Contractor

You will be aware of the interdependence of our works with those of your artist/independent contractor/statutory authority carrying out the installation of .... and that in accordance with the requirements of Clause 50 of the conditions we have afforded them every reasonable facility in the execution of their works on site.

Their works however are now behind programme/[and] greatly increased in scope and despite all reasonable efforts that we have made to integrate their operation into our overall programme, we find that in the area of .... our works are being seriously disrupted and impeded by ....

Having no privity of contract with the organisation concerned through which we can seek remedy, we now consider such firm to be causing us properly and directly to incur expense through prolongation and disruption beyond that provided for in or reasonably contemplated by the contract, and immediately upon becoming aware of the matter give notice pursuant to Clause 53(3)(c)(i) of the conditions requiring the reimbursement of such additional expense by way of increase in the contract sum.

Details of the costs incurred as required by Clause 53(3)(c)(ii) are appended/will be forwarded when available in substantiation of our application to be paid such sums by way of advance under Clause 40(5) of the conditions.

Yours faithfully

Contractor Limited

GCW1-M

Reference: **GC/50**

To: The SO

Dear Sir

Notice of Prolongation and Disruption Expenses

It is now apparent that compliance with SO's instructions concerning ..../making good of loss or damage to the works arising from an accepted risk under Clause 26(2)/delay in the provision of drawings/schedules/levels/the execution of work by others under the order of the Authority/nominating .... is unavoidably resulting in the regular progress of the works/part of the works being .... disrupted/[and]/prolonged causing the Contractor properly and directly to incur expense beyond that provided for in or reasonably contemplated by the Contract.

[This has occurred despite the fact that the contractor gave notice requesting the information in accordance with the requirements of Clause 53(3)(b).]

Immediately upon becoming aware of the matter we give notice pursuant to Clause 53(3)(c)(i) of the conditions requiring reimbursement of such additional expense by way of increase in the contract sum.

Details of the costs incurred as required by Clause 53(3)(c)(ii) are appended/will be forwarded when available in substantiation of our application to be paid such sums by way of advance under Clause 40(5) of the Conditions.

Yours faithfully

Contractor Limited

Reference: **GC/51**

To: The SO

Dear Sir

Request for Additional Information

Examination of the various drawings, specifications and instructions currently in our possession leaves us in some doubt as to the details/ [and]/specification for ....

We therefore request further drawings/an instruction and give notice requiring the item by .... pursuant to Clause 53(3)(b) of the conditions.

We believe this date is neither unreasonably early nor late in relation to the time when the information is required [and merely confirms the date previously incorporated on the agreed programme].

Yours faithfully

Contractor Limited

Reference: **GC/SO/56**

To: The Contractor

Dear Sir

Provision of Facilities for Other Contractors

You are required to provide .... for the use of .... who will shortly commence work/is currently working on site. The firm is an independent contractor and communication with them should only be made through ourselves.

This direction is given pursuant to Clause 50 of the Conditions and the nature of the requirement noted was specifically indicated [and priced] in the bill of quantities/shall be deemed to be a variation/is considered to be a requirement reasonably to be foreseen by an experienced contractor at time of tender.

Yours faithfully

Superintending Officer

Reference: **GC/SO/57**

To: The Contractor

Dear Sir

Acknowledgement/Rebuttal of Submission that Provision of Facilities for Other Contractors Could not Reasonably Have Been Foreseen

We refer to your letter dated .... reference .... giving notice that you consider that the provision of facilities for .... could not reasonably have been foreseen by an experienced contractor at the time of tender.

[We do not accept the validity of your notice given pursuant to Clause 50 of the conditions, since:

    [( ) the nature of the facilities required was specifically indicated in the tender documents and we draw your attention to ....]

    [( ) the facilities required are normally called for in the industry and should have been readily foreseeable by an experienced contractor]

    [( ) ....]]

[We accept the validity of your notice given pursuant to Clause 50 of the Conditions and note that you anticipate suffering delay/[and]/additional cost in consequence.]

[Details of your additional costs should be promptly submitted in accordance with the requirements of the Contract.]

[We note that any such delay is [not] critical to the timely construction of the overall works/liable to delay section .... of the works.]

Yours faithfully

Superintending Officer

CHAPTER 19

Security Arrangements and Arbitration

Clauses 55, 56, 57, 58 and 59—Corruption and Security Provisions

These clauses deal with the inter-related subjects of corruption, site admission, passes, photographs and secrecy.

Comment

1. Under Clause 55 the making of a gift or the payment of any form of commission in connection with the procurement of a contract is strictly forbidden. In the event of a corrupt gift being made or commission given, the contract may be terminated and the monetary value of the gift or commission recovered from the contractor. The decision of the Authority on such a matter shall be final and conclusive.

2. Under Clause 56 the following is a synopsis of the Authority's powers:

    (a) The Authority may give the contractor a notice that a person is not to be admitted to the site.

    (b) Aliens may be debarred from the site unless written permission has been given for their admission by the Authority.

    (c) The contractor may be instructed to furnish a list of names and addresses to the SO of all persons associated with the works.

    (d) The decision of the Authority as to whether a person is to be admitted to the site or not shall be final.

3. Under Clause 57 passes may need to be issued to gain access to the works.

Where the need for passes has not been highlighted in the tender enquiry documents and is introduced by way of variation, the contractor may well have ground for claim for additional expense under Clause 9(2) and possibly even under Clause 53.

4. Under Clause 58 the contractor is not allowed to take photographs unless he shall have first obtained the permission in writing of the Authority.

As a matter of normal, commercial routine such permission should be obtained, as no better form of evidence in dispute exists than a dated photographic record of work progress.

5. Under Clause 59 the contractor may be required to comply with the requirements of the Official Secrets Acts 1911 to 1939 and where

appropriate with the provisions of Section 11 of the Atomic Energy Act
1946.

Clause 61—Settlement of Disputes and Arbitration

(1) All disputes, differences or questions between the parties to the Contract with
respect to any matter or thing arising out of or relating to the Contract other than
a matter or thing arising out of or relating to Condition 51 or as to which the
decision or report of the Authority or of any other person is by the Contract
expressed to be final and conclusive shall after notice by either party to the
Contract to the other of them be referred to a single Arbitrator agreed for that
purpose, or in default of such agreement to be appointed at the request of the
Authority by the President of such one of the undermentioned as the Authority
may decide viz, the Law Society (or when, appropriate, the Law Society of
Scotland), the Royal Institute of British Architects, the Royal Institution of
Chartered Surveyors, the Royal Incorporation of Architects in Scotland, the
Institutions of Civil Engineers, Mechanical Engineers, Heating and Ventilating
Engineers, Electrical Engineers or Structural Engineers.

(2) Unless the parties otherwise agree, such reference shall not take place until
after the completion, alleged completion or abandonment of the Works or the
determination of the Contract.

(3) In the case of the Contract being subject to English Law such reference shall
be deemed to be a submission to arbitration under the Arbitration Act 1950, or
any statutory modification or re-enactment thereof.

(4) In the case of the Contract being subject to Scots Law the Law of Scotland
shall apply to the arbitration and the award of the Arbiter, including any award
as to the amount of any compensation, damages and expenses to or against any
of the parties to the arbitration, shall be final and binding on the parties, provided
that at any stage of the arbitration the Arbiter may, and if so requested by either
of the parties shall, prepare a statement of facts in a special case for the opinion
and judgement of the Court of Session on any question or questions of Law
arising in the arbitration, and both parties to the arbitration shall be bound to
concur in presenting to the Court a special case in the terms prepared by the
Arbiter and in which the statement of facts prepared by him is agreed by the
parties with such contentions as the parties or either of them may desire to add
thereto for the opinion and judgement of the Court; and the Arbiter and the
parties to the arbitration shall be bound by the answer or answers returned by the
Court of Session, or if the case is appealed to the House of Lords, by the House,
to the question or questions of Law stated in the case.

Comment

1. It will be seen from the clause that there is no lead up to the formal
disputes procedure comparable to that of a request for an engineer's
written decision as occurs under the ICE Conditions of Contract.

2. In the event of the parties failing to agree on an arbitrator it is the
Authority that will make the request to the President of the appropriate
professional institution. The wide-ranging scope of work on which these
conditions are used can be appreciated from the extremely comprehensive
schedule of professional institutions laid down.

3. For all practical purposes arbitration will not take place until after the alleged completion or abandonment of the works or determination of the contract, since it is unlikely that an Authority that finds difficulty in settling normal claims within the time-scale of the Contract would be prepared to resort to the far more fundamental procedure of arbitration in that time.

4. At various points in the conditions it will have been noted that the decision of the SO or the Authority has been stated to be final and conclusive. In these cases the matters stipulated are outside the authority of the arbitrator. In general it is not matters of crucial commercial significance that are so debarred, as will be seen from the schedule set out below:

(a) *Clause 3(2)*—sanction of the SO is required before items can be removed from the site and the SO's decision is final and conclusive.

(b) *Clause 7(3)*—the extent to which instructions are necessary or expedient shall be within the final and conclusive decision of the SO.

(c) *Clause 9(3)*—where works have been covered up without the SO having been given notice under Clause 22 prior to covering up work, the quantity surveyor shall be entitled to appraise the value and his decision shall be final and conclusive.

(d) *Clause 13(3)*—where an independent expert is called upon to carry out tests on materials prior to their incorporation in the works his decision shall be final and conclusive.

(e) *Clause 36(1) and (3)*—the SO may require the contractor to remove from the site any person of foreman status and below and the decision of the Authority or the SO shall be final and conclusive.

(f) *Clause 36(2) and (3)*—the Authority may require the contractor to remove from the site any person above foreman status and the decision of the Authority shall be final and conclusive.

(g) *Clause 40(6)*—the decision of the SO as to whether an amount has been properly paid and the amount to be paid to a nominated sub-contractor shall be final and conclusive.

(h) *Clause 42(3)*—the decision of the SO as to the amount to be paid on a fortnightly interim certificate shall be final and conclusive.

(i) *Clause 42(3)*—the decision of the Authority as to the amount of a normal monthly interim certificate shall be final and conclusive.

(j) *Clause 42(3)*—the decision of the Authority as to the issue of a completion certificate shall be final and conclusive.

(k) *Clause 44(5)*—the decision of the Authority as to reimbursement under a hardship plea following a special determination under

Clause 44 shall be final and conclusive. (Of the grounds for conclusive decision this appears the hardest and on the face of it a singularly unfair area of unilateral decision).

(l) *Clause 44(2) (a) (vi)*—the decision of the Authority as to instructions that need to be issued following determination under the special powers of determination shall be final and conclusive.

(m) *Clause 45 (c)*—the decision of the Authority as to a determination of the contract due to a default of the contractor, where the default arises under Clause 56 and the contractor has allowed access to the site of a person debarred by the Authority, shall be final and conclusive.

(n) *Clause 55(3)*—the decision of the Authority as to any matter arising under Clause 55 relating to corrupt practices shall be final and conclusive.

(o) *Clause 56(4)*—the decision of the Authority as to admissability of persons to the site shall be final and conclusive.

5. The need for the substantial sub-section (4) in the section is because the Arbitration Act 1950 does not apply to Scotland and the consequent need to legislate for matters which in England would have been covered by that Act.

STANDARD LETTERS

Reference: **GC/52**

To: The Authority

Dear Sir

Request for Permission to take Progress Photographs of the Works

We request your written permission pursuant to Clause 58 of the conditions to take photographs of the works for the purpose of maintaining an accurate progress record.

Yours faithfully

Contractor Limited

Reference: **GC/53**

To: The Authority

Dear Sir

Request to Concur in the Appointment of an Arbitrator

We refer to the dispute/difference between the contracting parties concerning .... and now give written notice requesting your concurrence in the appointment of an arbitrator pursuant to Clause 61(1) of the conditions.

We would suggest that immediate application be made to the President for the time being of ....
....
.... for the appointment of a suitable arbitrator.

Yours faithfully

Contractor Limited

Reference: **GC/SO/58**

To: The Contractor

Dear Sir

Refusal of Admission to the Site

We hereby inform you that the undernoted person/persons is/are to be refused admission to the Site pursuant to Clause 56(1) of the Conditions.

[..]
[..]

[Appreciating this to be a matter of delicacy we have refrained from stating our detailed reasons for our directive in this notice but will be prepared to discuss the matter with a senior and accredited representative from your company.]

Yours faithfully

Superintending Officer

Reference: **GC/SO/59**

To: The Contractor

Dear Sir

List of Contractor's Site Personnel

We hereby request that you furnish the details of the names, addresses, capacity in which employed [and ....] of all site personnel and others who are or may be at any time concerned with the works.

This request is made pursuant to Clause 56(3) of the Conditions.

Yours faithfully

Superintending Officer

Reference: **GC/SO/60**

To: The Contractor

Dear Sir

Consent to take Progress Photographs of the Site/Works

We refer to your request to take progress photographs of the site/works and confirm our consent pursuant to Clause 58 of the Conditions.

Yours faithfully

Superintending Officer

Reference: **GC/SO/61**

To: The Contractor

Dear Sir

Notification to Concur in the Appointment of an Arbitrator

We refer to the dispute/difference between the contracting parties concerning .... and now give written notice requesting your concurrence in the appointment of an arbitrator pursuant to Clause 61(1) of the conditions.

We would suggest that immediate application be made to the President for the time being of .... for the appointment of a suitable arbitrator.

Yours faithfully

Superintending Officer

Index